T0073159

Quantum Mechanics for Tomorrow's Engineers

Discover the foundations of quantum mechanics, and explore how these principles are powering a new generation of advances in quantum engineering, in this ground-breaking undergraduate textbook. It explains physical and mathematical principles using cutting-edge electronic, optoelectronic and photonic devices, linking underlying theory with real-world applications; focuses on current technologies and avoids historic approaches, getting students quickly up-to-speed to tackle contemporary engineering challenges; provides an introduction to the foundations of quantum information, and a wealth of real-world quantum examples, including quantum well infrared photodetectors, solar cells, quantum teleportation, quantum computing, band gap engineering, quantum cascade lasers, low-dimensional materials, and van der Waals heterostructures; and includes pedagogical features such as objectives and end-of-chapter homework problems to consolidate student understanding, and solutions for instructors. Designed to inspire the development of future quantum devices and systems, this is the perfect introduction to quantum mechanics for undergraduate electrical engineers and materials scientists.

JUNICHIRO KONO is Karl F. Hasselmann Chair in Engineering, Professor of Electrical & Computer Engineering, Professor of Physics & Astronomy, Professor of Materials Science & NanoEngineering, and Chair of Applied Physics at Rice University, which he joined in 2000. His research focuses on optical studies of condensed matter systems and photonic applications of nanosystems, including semiconductor nanostructures and carbon-based nanomaterials. He has made a number of pioneering contributions to the diverse fields of semiconductor optics, terahertz spectroscopy and devices, ultrafast and quantum optics, and condensed matter physics. He is a Fellow of APS, OSA, and SPIE. He is a leader in optical studies of condensed matter systems and photonic applications of nanosystems.

"Kono's ten years of teaching a course to engineering undergraduates leads to an introductory book on quantum and information science with refreshing directness, which may also serve as a reference. It includes clear concepts of quantum entanglement and measurement, information theory of circuits and algorithms, and solid accounts of the quantum advantage and quantum materials"

L. J. Sham, University of California, San Diego

"This book provides the essential shortcuts for beginners to learn quantum physics that can be directly applied to the frontline research in modern quantum engineering applications. The author makes an excellent effort to balance rigor and intuition by selecting suitable pedagogical topics, while preserving the core of quantum mechanics for engineering science."

Philip Kim, Harvard University

JUNICHIRO KONO
Rice University, Houston

QUANTUM MECHANICS FOR TOMORROW'S ENGINEERS

CAMBRIDGE
UNIVERSITY PRESS

CAMBRIDGE
UNIVERSITY PRESS

University Printing House, Cambridge CB2 8BS, United Kingdom

One Liberty Plaza, 20th Floor, New York, NY 10006, USA

477 Williamstown Road, Port Melbourne, VIC 3207, Australia

314–321, 3rd Floor, Plot 3, Splendor Forum, Jasola District Centre,
New Delhi – 110025, India

103 Penang Road, #05–06/07, Visioncrest Commercial, Singapore 238467

Cambridge University Press is part of the University of Cambridge.

It furthers the University's mission by disseminating knowledge in the pursuit of
education, learning, and research at the highest international levels of excellence.

www.cambridge.org
Information on this title: http://www.cambridge.org/highereducation/
isbn/9781108842587
DOI: 10.1017/9781108903592

First published 2023

A catalogue record for this publication is available from the British Library.

Library of Congress Cataloging-in-Publication Data

ISBN 978-1-108-84258-7 Hardback

Additional resources for this publication at www.cambridge.org/Kono.

Contents

To Alissa, Renée, and Yuko

Preface

This book is for engineers and material scientists interested in learning quantum mechanics. Though many readers will read this book out of intrinsic interest in the subject, others will do so out of necessity: they will find their colleagues interpreting data, diagnosing problems with instruments, or designing better devices using quantum mechanics. Perhaps there was a time when engineers and materials scientists were better off investing their time in other areas of study, but that time is over. There seems to be no technical field that quantum mechanics has not invaded.

Quantum mechanics was born and developed near the beginning of the twentieth century, and it has become one of today's most fundamental physical theories. Its principles provide the most powerful and precise ways of describing the universe – from the smallest subatomic particles to the large-scale structure of the cosmos. The methods and implications of quantum mechanics have not only profoundly influenced how scientists and philosophers think and look at nature but also have significantly contributed to the development of today's technology in computation and communications.

It is evident that the impact of quantum mechanics will continue to increase in electronics, optoelectronics, and photonics, and will further expand into many other technological fields, including energy, artificial intelligence, thermoelectrics, medicine, biotechnology, and neuroengineering. Most importantly, the degree of miniaturization of modern electronic devices has reached the point where quantum mechanical effects have to be explicitly considered and precisely controlled. In the not-so-distant future, Moore's law will cease to apply, and entirely new, genuinely quantum technologies will need to be developed to go beyond today's silicon-based classical solid-state device technologies.

Every undergraduate student, not only in science but also in engineering, must thus study quantum mechanics. Although there are many excellent volumes of undergraduate-level quantum mechanics textbooks, only a few are geared toward engineering students.

Book Outline

Specifically, the goal of this book is to provide engineering undergraduate students with the physical understanding and mathematical prowess necessary for designing quantum devices and systems, including quantum computers. The main players throughout this book are thus electrons in solids (as opposed to electrons in atomic gases, plasmas, or biological cells).

By studying this textbook, it is hoped that students will learn: (1) when, or under what conditions, quantum effects become non-negligible in devices, (2) how to calculate the quantum states (wavefunctions) and energies of electrons moving in artificial potential structures, and (3) what determines the electrical and optical properties of quantum solid-state devices.

This book foregoes the standard introductory quantum mechanics textbook approach of introducing quantum mechanics through a historical series of crises in late nineteenth century physics. Instead, quantum mechanics is introduced as a theory that is in use today in a wide variety of electronic, optoelectronic, and photonic devices. The essential mathematical foundations of quantum mechanics are introduced as early as possible and then immediately followed by an introduction to quantum information science, which is one of the most significant emerging applications of quantum mechanics.

Among topics that this book does *not* cover are: the hydrogen atom; quantum statistics; scattering; angular momentum; symmetry and transformations; and variational approaches. These topics were not included since they are fairly advanced and of limited use for an engineering student.

Overall, the book aims to strike a balance between rigorous mathematical derivations and example problem solving. The example problems are highly practical, including such modern topics as quantum well infrared photodetectors, solar cells, quantum teleportation, band gap engineering for LEDs, quantum cascade lasers, and band diagrams of van der Waals heterostructures.

How This Book Can Be Used for Teaching

The book is primarily aimed at undergraduate students in electrical engineering and materials science, and possibly in computer science, mechanical engineering, and chemical engineering, who are interested in learning quantum mechanics. In addition, the book can be useful for professional engineers working in areas of quantum devices, quantum computation, and quantum communications.

This text can be used for a one-semester course. It is based on the lecture notes that the author developed for his course ELEC 361: "Quantum Mechanics for Engineers," which he taught for a total period of eleven years. This was a core course in the photonics, electronics, and nanoelectronics specialization of the electrical and computer engineering curriculum at Rice University, intended as an introduction to the subject for electrical and computer engineering majors, typically in their junior or senior year.

The reader is expected to have a prior knowledge of differential equations and linear algebra. Online Resources for Instructors include: a Solutions Manual, PowerPoint lecture slides, and PowerPoint slides and JPEGs of all the figures and tables from the book.

Bibliographical References and Sources

There are a number of well-regarded undergraduate-level quantum mechanics textbooks for physics students, including

- R. P. Feynman, R. B. Leighton, and M. Sands, *The Feynman Lectures on Physics, Vol. 3: Quantum Mechanics* (Addison Wesley, 1964).

- R. Eisberg and R. Resnick, *Quantum Physics of Atoms, Molecules, Solids, Nuclei, and Particles, Second Edition* (John Wiley & Sons, 1985).

- S. Gasiorowicz, *Quantum Physics, Third Edition* (John Wiley & Sons, 2003).

- J. S. Townsend, *A Modern Approach to Quantum Mechanics, Second Edition* (University Science Book, 2013).

- D. J. Griffiths and D. F. Schroeter, *Introduction to Quantum Mechanics, Third Edition* (Cambridge University Press, 2018).

The following excellent textbooks, on the other hand, are geared toward undergraduate students studying engineering and applied physics, and are written in a similar spirit to the current textbook:

- A. Yariv, *An Introduction to Theory and Applications of Quantum Mechanics* (John Wiley & Sons, 1982).

- H. Kroemer, *Quantum Mechanics for Engineering: Materials Science and Applied Physics* (Pearson, 1994).

- J. Singh, *Quantum Mechanics: Fundamentals and Applications to Technology* (John Wiley & Sons, 1996).

- D. K. Ferry, *Quantum Mechanics: An Introduction for Device Physicists and Electrical Engineers, Second Edition* (Taylor & Francis, 2001).

- D. A. B. Miller, *Quantum Mechanics for Scientists and Engineers* (Cambridge University Press, 2008).

- A. F. J. Levi, *Applied Quantum Mechanics, Second Edition* (Cambridge University Press, 2012).

Furthermore, because of the focus of the present textbook on electrons in solid state devices, the following introductory textbooks on solid state physics, semiconductor physics, and materials science should be useful:

- J. H. Davies, *The Physics of Low-Dimensional Semiconductors: An Introduction* (Cambridge University Press, 1997).

- J. D. Livingston, *Electronic Properties of Engineering Materials* (John Wiley & Sons, 1999).

- D. A. Neamen, *Semiconductor Physics and Devices, Fourth Edition* (McGraw-Hill, 2012).

- S. H. Simon, *The Oxford Solid State Basics* (Oxford University Press, 2013).

- S. O. Kasap, *Principles of Electronic Materials and Devices* (McGraw-Hill, 2018).

Acknowledgments

I am grateful to all Rice University students who served as graders and/or teaching assistants for the ELEC 361 course over the years – Lei Ren, Erik H. Hároz, Thomas A. Searles, Darius T. Morris, Layla G. Booshehri, G. Timothy Noe II, Qi Zhang, Minjie Wang, Nick A. Thompson, Weilu Gao, Cody Sewell, Ahmed Zubair, Xinwei Li, Natsumi Komatsu, Yakub Grzesik, Dasom Kim, and Elijah Kritzell. In particular, Nick Thompson provided crucial help in developing an initial draft of this textbook, and Elijah Kritzell prepared solutions to the exercise problems. Finally, I thank Deyin Kong for creating many artistic figures used in this textbook, Dr. Palash Bharadwaj for useful discussions on various subjects covered in the book, and Andrey Baydin, Nolan Bitner, Nicolas Marquez Peraca, Fuyang Tay, and Hongjing Xu for proofreading the entire manuscript.

Junichiro Kono

1

Quantum-Enabled Technologies

THE FIELD OF QUANTUM RESEARCH is currently undergoing a revolution. A variety of tools and platforms for controlling individual quantum particles have emerged, which can be utilized to develop entirely new technologies for computation, communication, and sensing. In particular, these technologies will enable applications of quantum information science that can fundamentally change the way we store, process, and transmit information. Exciting theoretical predictions exist for quantum computers, with some proof-of-principle experiments, to perform calculations that would overwhelm the world's best conventional supercomputers. Quantum research is rapidly developing, and the race is intensifying for quantum technology development, involving some of the high-tech giants. In this chapter we will introduce some key concepts in the materials and devices behind these technological developments. Becoming familiar with these concepts in this first chapter should provide the reader with concrete goals and motivations for studying the quantum methods and tools described in subsequent chapters.

Learning objectives:

- Understanding the limit of miniaturization of electronic devices and the need for genuinely quantum devices.

- Becoming familiar with some of the key emerging concepts, materials, and devices that go beyond the traditional silicon-based technologies.

- Getting a glimpse of the currently occurring second quantum revolution, especially the advent of quantum information processing.

1.1 Limit of Miniaturization

Moore's law assumes that the typical feature size of a microelectronic circuit in computers shrinks exponentially as a function of time (see Figure 1.1). If the current trend were to persist, the typical feature size would soon approach the size of an atom, which is the ultimate end of this law. Even before this ultimate end is reached, the classical laws are already breaking down and being replaced by the laws of quantum physics. For instance, in nanostructures electrons behave like waves, obeying a new wave equation – the Schrödinger equation (see Chapter 2) – rather than Newton's equation of motion.[1] Electrons exhibit nonclassical phenomena – possessing discrete energies and

[1] See Appendix A.

Figure 1.1 Miniaturization trend for electronic devices. Gordon E. Moore, one of the founders of Intel, observed in 1965 that the number of transistors in an integrated circuit doubles about every two years. By around 2050, the size of a transistor will reach the size of an atom, if this trend continues. The inset image of cells was adapted from https://commons.wikimedia.org/wiki/File:Wilson1900Fig2.jpg.

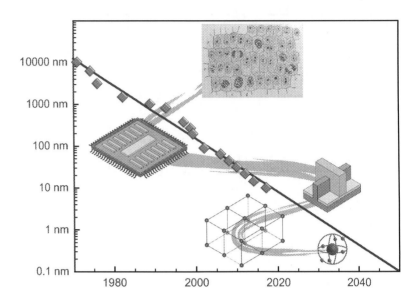

[2] *Some of the key nonintuitive quantum concepts and principles behind the quantum devices that are under development:*

- Wave–particle duality
- Zero-point energy
- Superposition
- Complementarity
- The uncertainty principle
- Wavefunction collapse
- Entanglement

tunneling through classically impenetrable barriers – which prevent classical mechanics-based devices from operating normally. Therefore, researchers in this field are searching for new ways of operating these devices for the further improvement of computation power by actively taking advantage of nonintuitive quantum concepts and principles.[2]

Quantum physics is thus believed to play a central role in the further advancement of information processing technology in the twenty-first century. It is true that quantum mechanics lies behind today's solid state devices already; without it, one cannot even explain, e.g., how the transistor works, why different semiconductor LEDs produce different colors, and what makes silicon unsuitable for laser diodes. In conventional semiconductor devices, however, quantum mechanics is not always needed once one accepts concepts like band structure, effective mass, and Fermi energy. One can safely use Newton's equation of motion for "effective-mass particles" (together with some statistical mechanics ideas) in most situations in order to understand how semiconductor devices operate. In the genuinely quantum devices that researchers are currently developing, on the other hand, quantum concepts will be more actively used, and they are expected to lead to not only better performance but also entirely new functionalities or multifunctionalities. These revolutionary ideas will likely be implemented in devices made of recently discovered materials – e.g., nanomaterials and quantum materials – with unusual properties and capabilities beyond those of well-developed traditional materials such as silicon (Si) and gallium arsenide (GaAs).

1.2 *Spin Instead of Charge*

A classical particle in a three-dimensional (3D) world has three or-bital degrees of freedom along, say, the x-, y-, and z- directions. Similarly, an electron in a conventional semiconductor device has three orbital degrees of freedom. If the semiconductor device is very thin in the z-direction (so thin that its thickness is comparable with or smaller than the quantum mechanical wavelength of the electrons), electrons in the device move two-dimensionally (2D), having only two degrees of freedom, (x, y), since they are *quantum-confined* in the z-direction; see Section 2.2. In a wire-shaped, ultranarrow semiconductor device (which can be fabricated using modern device processing technology), electrons are confined in a one-dimensional (1D) world, possessing just one degree of freedom (say, x). Since electrons are charged, they move in response to an electric field, or a voltage, in the unconfined direction(s) in a given device. See Figure 1.2.

Quantum mechanically, however, electrons have a fourth degree of freedom, called spin, which has an associated magnetic moment.[3] This is an authentically quantum mechanical, internal degree of freedom, having two discrete states called spin "up" and spin "down"; see Figure 1.3. The operation mechanisms of traditional devices, however, are all charge- or orbital-based, where the electrons move in x, y, and/or z coordinates driven by an electric field; see Figure 1.2. There is much interest in developing spintronic devices. In spintronics, the spin degree of freedom of the electron is actively utilized in addition to, or in place of, the charge (orbital) degrees of freedom. Expected improvements to conventional charge-based devices include nonvolatility, increased data processing speed, decreased electric power consumption, and increased integration densities. Electron spins in semiconductors have been recognized as an ideal medium on which to encode quantum bits, or qubits,[4] owing to their long coherence lifetimes. It is also expected that electronic, magnetic, and photonic functions can be incorporated into single devices to create spin-based multifunctional devices.

The first generation of spintronic devices, based on metals, has already been commercially available in the form of read heads inside hard disks based on the phenomenon of giant magnetoresistance.[5] Semiconductor-based spintronic devices are expected to have a richer variety of capabilities and utilities because of the controllability of semiconductors. Further, the spin and orbital degrees of freedom can mix in semiconductors through spin–orbit coupling, so electron spins can respond not only to magnetic fields but also to electric fields. A number of different types of innovative semiconductor spintronic

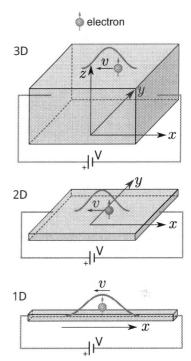

Figure 1.2 Semiconductors in different dimensions. © Deyin Kong (Rice University).

[3] Its magnitude in vacuum, called the Bohr magneton, is equal to $\mu_B = e\hbar/2m_e = 9.274009994 \times 10^{-24}$ J T^{-1}, where e is the electronic charge, $\hbar$ is the reduced Planck constant (see Section 2.1.1), and m_e is the electron mass in vacuum.

[4] See Section 4.1.

[5] Albert Fert and Peter Grünberg shared the 2007 Nobel Prize in Physics for their discovery of giant magnetoresistance, or GMR.

Figure 1.3 The two distinct spin states of an electron – the spin "up" and "down" states. © Deyin Kong (Rice University).

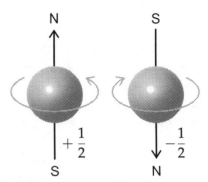

[6] See, e.g., S. A. Wolf *et al.*, *Science* **294**, 1488 (2001); D. D. Awschalom, D. Loss, and N. Samarth, Eds., *Semiconductor Spintronics and Quantum Computation* (Springer, 2002); and J.-B. Xia, K. Chang, and W. Ge, *Semiconductor Spintronics* (World Scientific, 2012).

devices have been proposed, and some of them have already been experimentally demonstrated or implemented.[6] These include spin resonant tunneling diodes (spin RTDs), spin field effect transistors (spin FETs), and spin light emitting diodes (spin LEDs). They are expected to bring new functionalities that conventional devices do not have. For example, a spin LED is a conversion device where information can be transferred from spin-polarized current (electronic spins) to light (photons); see Figure 1.4. The most challenging goals of spintronic devices are to create, manipulate, and detect *coherent superposition* states of spin, which can be used for quantum information processing (see Section 1.5 and Chapter 4).

Figure 1.4 Spin LED. A spin-polarized current is converted into a circularly polarized light beam. For a recent review on spin LEDs, see, e.g., N. Nishizawa and H. Munekata, *Proceedings of SPIE* **11090**, 1109034 (2019); DOI: 10.1117/12.2527862. © Deyin Kong (Rice University).

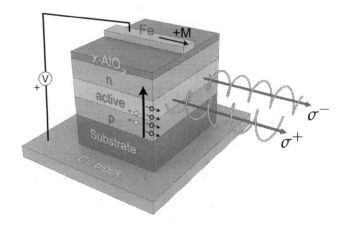

1.3 Photons Instead of Electrons

As detailed in Section 2.1.1, light consists of "bundles" of energy, called photons.[7] Photons are quantum particles and possess, just like electrons, an internal degree of freedom, called polarization, which is expressed in terms of the directions of the electric and magnetic

[7] Albert Einstein proposed the existence of such bundles in 1905, the year when he also published his ground-breaking paper on special relativity. The name *photon* was coined by Gilbert N. Lewis in 1926; see *Nature* **118**, 874 (1926).

fields of the light wave, as is familiar in the classical electromagnetism theory of Faraday and Maxwell.[8] Just like electrons' spins, photons' polarization states can be used to encode quantum information. One can express these polarization states either in a circular basis (with "right" and "left" basis states) or in a linear basis (with "horizontal" and "vertical" basis states). Any general polarization state can be expressed as a superposition (i.e., a linear combination) of basis states. See Chapter 3 for the basic properties of superposition states and Chapter 4 for how superposition states are used in quantum information processing.

Since photons are not charged particles, they cannot be accelerated or manipulated by an electric or magnetic field, and they normally do not interact with each other. These properties may seem to be drawbacks in certain applications, but insensitivity to external fields also means that they are insensitive to external perturbations, which leads to robust quantum properties, especially quantum coherence. This is a great advantage in quantum technology applications since quantum states are usually fragile. Once they are created, quantum superposition and/or entangled states (see Chapter 4) of photons can live for a long time. Furthermore, photons can be easily transported in free space or through an optic fiber, acting as "flying qubits" and transmitting quantum information over a long distance. For these reasons, photons play the dominant role as information carriers in quantum networks and communications (see Figure 1.5).

The fundamental quantum properties of light have been traditionally studied within the field of quantum optics, particularly when light is in interaction with atoms and molecules. However, such properties have recently begun to be exploited in photonic devices and systems based on solid-state materials. A variety of single-photon sources in different wavelength ranges have been built, and different types of transducers for converting an electronic qubit into a photon qubit are being produced. Efficient methods for enhancing photon-photon interactions and creating entangled photon states (Figure 1.6) have been developed for various media, and quantum teleportation has been demonstrated using entangled photons; see Section 4.5. Most recently, scientists have built a quantum information link between an orbiting craft and its terrestrial controllers using entangled photons.[9]

[8] See Appendix C.

Figure 1.5 Photons (small circles) can serve as fast and robust information carriers in quantum networks. © Deyin Kong (Rice University).

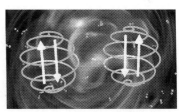

Figure 1.6 Entangled photons with mutually perpendicular polarization. © Deyin Kong (Rice University).

[9] J. Yin *et al.*, *Nature* **582**, 501 (2020)

1.4 Materials Beyond Silicon

Traditional electronic devices, such as diodes and transistors, are constructed from the elemental semiconductor Si, and traditional photonic devices, such as LEDs and laser diodes, are constructed from III-V compound semiconductors, including GaAs, indium phosphide (InP), and gallium nitride (GaN). These materials have been around for many decades, and their basic properties are well understood within the band theory of solids.[10] During the last decade, several new classes of materials have emerged, whose properties are either superb (compared with traditional materials) or truly unique in certain aspects. These materials – sometimes called quantum materials – exhibit properties or phenomena that cannot be understood using classical concepts alone.[11] Understanding these materials requires some of the concepts that we will explore in later chapters: quantum superposition, entanglement, and topology.

[10] See Chapter 7.

[11] For a review, see, e.g., "The rise of quantum materials," *Nature Physics* **12**, 105 (2016).

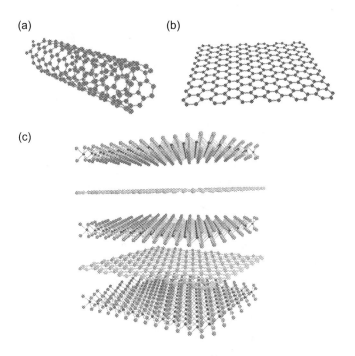

Figure 1.7 Nanomaterials: (a) carbon nanotube, (b) graphene, and (c) heterostructure of 2D materials including graphene and transition metal dichalchogenides. © Deyin Kong (Rice University).

[12] See Section 7.5 for details about graphene.

Nanomaterials are those materials in which electrons are quantum mechanically confined in at least one spatial dimension. For example, graphene[12] is a 2D conductor in which massless electrons can move rapidly, like photons, only in a one-atom-thick plane of covalently bonded carbon atoms. Similarly, atomically thin transition metal dichalcogenides have been realized that host strongly bound 2D electron–hole pairs, called 2D excitons, which govern the optical properties of these materials with an additional "valley degree"[13] of freedom. In a carbon nanotube, electrons are confined in two spatial dimensions, making it a 1D system, or a quantum wire. It should be emphasized that these materials are *naturally* low-dimensional, not as a result of nanofabrication as in the case of lithographically defined artificial semiconductor quantum structures.[14]

In topological insulators (TIs), the spin and orbital degrees of freedom are locked, which creates an unusual flow of spin-polarized electrons propagating in a certain direction specified by the spin orientation, i.e., at 2D surfaces in a 3D material or 1D edges in a 2D material; see Figure 1.8. The quantum states of these electrons exhibit robust coherence, that is, they are insensitive to sample details, especially imperfections, since their spin and orbital states and dynamics are topologically protected.[15] Such robust quantum states are promising for creating quantum devices that work at high temperatures. Finally, there is a class of materials referred to as strongly correlated materials,[16] which exhibit behaviors beyond the conventional band theory[17] of solids – e.g., unusually high superconducting transition temperatures or unrealistically large effective masses. All these modern materials are currently under intense investigations for potential use in future quantum devices.

[13] See, e.g., J. R. Schaibley *et al.*, "Valleytronics in 2D materials," *Nature Reviews Materials* **1**, 16055 (2016).

[14] See Section 7.4.

[15] For an introductory review on topological insulators, see, e.g., J. E. Moore, *Nature* **464**, 194 (2010), and C. L. Kane and J. E. Moore, *Physics World* **24**, 32 (2011).

[16] For a review, see, e.g., E. Morosan, D. Natelson, A. H. Nevidomskyy, and Q. Si, *Advanced Materials* **24**, 4896 (2012).

[17] See Chapter 7.

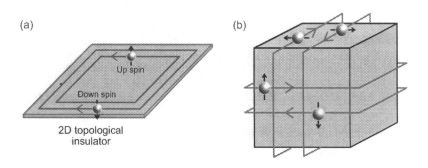

(a)   (b)

Up spin

Down spin

2D topological insulator

Figure 1.8 Edge and surface states of topological insulators. (a) 1D helical edge state of a 2D TI. (b) 2D helical surface states of a 3D TI. © Deyin Kong (Rice University).

1.5   Quantum Strategies for Information Processing

The rapid scientific and technological developments that we are currently witnessing can be viewed as the *second* quantum revolution. The first quantum revolution occurred during the first few decades of the twentieth century following the foundation of quantum mechanics, which drastically changed the way in which we describe the universe. Quantum theory provided new rules that explain physical processes in atoms, molecules, and solids. Scientists then used those rules to build revolutionary devices such as lasers and transistors, on which the information age was built. The second quantum

[18] J. P. Dowling and G. J. Milburn, "Quantum technology: The second quantum revolution," *Philosophical Transactions of the Royal Society of London A* **361**, 1655 (2003).

[19] See Chapter 4.

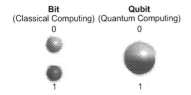

Bit   **Qubit**
(Classical Computing) (Quantum Computing)
0   0

1   1

Figure 1.9 While classical bits are either 1 or 0, quantum bits (qubits) can be in a superposition state, a linear combination, of 1 and 0. See Section 4.1 for more details. © Deyin Kong (Rice University).

[20] S. Lloyd, *Science* **273**, 1073 (1996). See also the paragraph on quantum simulation in Section 4.6.3.

[21] See Section 4.6.

[22] See Section 4.6.3.

revolution[18] is about controlling individual quantum systems to a much greater extent than before, enabling even more powerful applications of quantum mechanics. Many of these new applications rely on genuinely quantum, nonintuitive concepts such as superposition and entanglement. These concepts are becoming more and more common and important in diverse scientific disciplines beyond physics, including materials science, electrical engineering, chemistry, mathematics, and computer science. Together, researchers in these fields are developing novel quantum materials, devices, and systems, which can lead to practical applications of quantum technology.

In particular, quantum information science and technology (QIST) is rapidly expanding.[19] It is a multidisciplinary field of research on information storage, processing, and transmission based on the principles of quantum mechanics. Once developed, QIST devices and architectures are expected to disrupt nearly every sector of industry – e.g., computers, online security, cryptography, drug development, financial modeling, and weather forecasting. It is expected that QIST will help us fight disease, invent new materials, and solve health and climate problems. In a QIST system, quantum information is stored in the form of quantum bits (called qubits). Unlike classical bits, which can be either 1 or 0, qubits can be in both 1 and 0 simultaneously through quantum superposition, that is, a linear combination of 1 and 0; see Figure 1.9. Qubits are processed inside a quantum computer through quantum gates, and sent over long distances within a quantum network. If general-purpose, or universal,[20] fault-tolerant quantum computers are successfully built, they will revolutionize computation, being able to solve certain problems faster (sometimes *exponentially* faster) than any of today's fastest supercomputers.[21]

While the development of quantum computers and networks may require continuing efforts for the next few decades, many technologies, referred to as enabling (or bridging) technologies, do not require fully fledged quantum computers for demonstrating the so-called quantum advantages over classical computers.[22] For developing any of these applications, it is fundamentally important to construct quantum platforms and architectures that have robust quantum states. We have to be able to prepare a qubit whose quantum coherence can last for a long enough time for a significant number of quantum operations to be completed before decoherence occurs. Atomic, ionic, and molecular systems have the advantages of long coherence times, and they have been used for a number of proof-of-principle experiments, especially at ultralow temperatures, but they are difficult to scale up to a large quantum computer. Quantum materials possess some intrinsically quantum properties, arising from many-body interactions, which may provide robust, solid-state qubit systems.

1.6   Chapter Summary

In this chapter, we provided a brief overview of the cutting-edge research on technological applications of quantum mechanics. The exponential miniaturization trend of semiconductor microdevices and integrated-circuit components that has persisted in the past several decades, called Moore's law, is approaching the absolute physical limit (see Figure 1.1), and revolutionary ideas are needed to continue the current rate of increase in computation speeds. Quantum mechanics is believed to hold the key for further development of information processing in the twenty-first century, and we are currently experiencing a quantum revolution. The intrinsically quantum mechanical spin degree of freedom of electrons is actively utilized in spintronic devices, in addition to, or in place of, the usual orbital degrees of freedom. Photon-based quantum devices and technologies are being developed for securely storing and sending information, and a variety of new materials with extraordinary quantum properties have been discovered. All these rapid advances in basic research are stimulating much interest in quantum technology, especially in quantum information science and technology. Having browsed through these frontier topics in quantum technology, we are now ready to start delving into the detailed methods and ideas of quantum mechanics, especially those that are essential for tomorrow's engineers.

1.7 *Exercises*

The exercises below involve some of the basic mathematics as well as the basic concepts and methods of classical mechanics and electromagnetism that are used throughout the book, rather than the main subject of this chapter. For more details on classical mechanics, i.e., the Newtonian, Lagrangian, and Hamiltonian formulations of classical mechanics, see Appendix A, which includes additional exercises. For the classical theory of waves see Appendix B and, for the basics of Maxwell's theory of electromagnetism, see Appendix C.

Exercise 1.1 (Euler's formula)

Use Euler's formula to prove the following addition theorems:

$$\sin(x+y) = \sin x \cos y + \cos x \sin y, \tag{1.1}$$

$$\cos(x+y) = \cos x \cos y - \sin x \sin y. \tag{1.2}$$

Exercise 1.2 (Partial derivatives)

When

$$f(x,y,z) = \frac{1}{\sqrt{x^2 + y^2 + z^2}}, \tag{1.3}$$

show that $\nabla^2 f = 0$, where ∇^2 is the Laplacian operator.

Exercise 1.3 (Vector calculus)

In a vacuum, Maxwell's equations are written as $\nabla \cdot \boldsymbol{\mathcal{E}} = \rho/\varepsilon_0$, $\nabla \cdot \boldsymbol{B} = 0$, $\nabla \times \boldsymbol{\mathcal{E}} = -\partial \boldsymbol{B}/\partial t$, and $\nabla \times \boldsymbol{B} = (1/c^2)\partial \boldsymbol{\mathcal{E}}/\partial t$, where $\boldsymbol{\mathcal{E}} = -\nabla \phi - \partial \boldsymbol{A}/\partial t$ is the electric field vector, $\boldsymbol{B} = \nabla \times \boldsymbol{A}$ is the magnetic field vector, ϕ and $\boldsymbol{A}$ are the scalar and vector potentials, respectively, ρ is the charge density (assumed to be a constant here), ε_0 is the vacuum permittivity, and c is the speed of light. Using the Lorentz condition $\nabla \cdot \boldsymbol{A} + (1/c^2)\partial \phi/\partial t = 0$, derive the following equations:

$$\nabla^2 \phi - \frac{1}{c^2}\frac{\partial^2 \phi}{\partial t^2} = -\frac{\rho}{\varepsilon_0} \tag{1.4}$$

$$\nabla^2 \boldsymbol{A} = \frac{1}{c^2}\frac{\partial^2 \boldsymbol{A}}{\partial t^2} \tag{1.5}$$

Exercise 1.4 (Differential equations)

A particle of mass m (> 0) connected to a spring with spring constant K (> 0) obeys Newton's equation of motion:

$$-Kx = m\frac{d^2x}{dt^2}. \tag{1.6}$$

Here, $x(t)$ is the position of the particle measured from its equilibrium position at time t.

(a) Set up and solve the characteristic equation of this second-order differential equation. Namely, assume $x(t) = e^{\lambda t}$ and obtain the roots λ_1 and λ_2.

(b) What is the general solution to Equation (1.6)?

(c) What is the specific solution to Equation (1.6) that satisfies the initial conditions $x(0) = x_0$ and $\dot{x}(0) = v_0$?

Exercise 1.5 (Wave motion analysis)

Given the wavefunctions

$$\psi_1 = 4\sin 2\pi(0.2x - 3t), \tag{1.7}$$

$$\psi_2 = \frac{\sin(7x + 3.5t)}{2.5} \tag{1.8}$$

determine in each case the values of (a) frequency, (b) wavelength, (c) period, (d) amplitude, (e) phase velocity, and (f) direction of motion. Time t is given in seconds, and x is given in meters.

Exercise 1.6 (Three-dimensional wave equation)

The electric field of an electromagnetic wave in vacuum traveling in the positive x-direction is given by

$$\mathcal{E} = \mathcal{E}_0 e_y \sin\left(\frac{\pi z}{z_0}\right) \cos(kx - \omega t) \tag{1.9}$$

where e_y is a unit vector in the y-direction and z_0 is a constant.

(a) Using the wave equation, obtain an expression for $k = k(\omega)$.

(b) Find the phase velocity of the wave.

Exercise 1.7 (Wavepacket)

Let us take the superposition of an infinite number of plane waves

$$\psi(x,t) = \int_{-\infty}^{\infty} A(k) e^{-i(\omega t - kx)} dk \tag{1.10}$$

where $A(k)$ represents a spectral distribution function.

(a) When $A(k) = B\exp[-\sigma^2(k - k_0)^2]$, where $\sigma > 0$, calculate $\psi(x, 0)$. The following formula should be useful:

$$\int_{-\infty}^{\infty} e^{-ax^2} dx = \sqrt{\frac{\pi}{a}} \tag{1.11}$$

(b) Express the $1/e$ width of the wavepacket both in real space and k-space as Δx and Δk, respectively, in terms of σ. Show that $\Delta x \cdot \Delta k$ is constant and discuss the implications of this result.

Exercise 1.8 (Plasma reflection)

Let us consider the wave equation for an electric field $\mathcal{E}$ in a non-magnetic medium with conductivity σ and permittivity $\varepsilon = \varepsilon_r \varepsilon_0$:[23]

[23] See Equation (B.1) in Appendix B.

$$\nabla^2 \mathcal{E} = \mu_0 \sigma \frac{\partial \mathcal{E}}{\partial t} + \mu_0 \varepsilon \frac{\partial^2 \mathcal{E}}{\partial t^2} \tag{1.12}$$

where μ_0 is the vacuum permeability, ε_0 is the vacuum permittivity, and ε_r is the relative dielectric constant (assumed to be real and independent of frequency).

[24] See Appendix C.

(a) Derive Equation (1.12) from Maxwell's equations.[24]

(b) Assume that the field $\mathcal{E}$ varies harmonically both in time (t) and space (r), i.e., $\mathcal{E} = \mathcal{E}_0 \exp\left[-i(\omega t - \mathbf{k} \cdot \mathbf{r})\right]$, where ω is the angular frequency and $\mathbf{k}$ is the wavevector. Show that $k = |\mathbf{k}| = \tilde{N}\omega/c$, where

$$\tilde{N} = \left(\varepsilon_r + i \frac{\sigma}{\varepsilon_0 \omega} \right)^{1/2} \tag{1.13}$$

is the complex refractive index and c is the speed of light.

[25] See Figure C.1 in Appendix C.

(c) Use the AC Drude conductivity formula[25]

$$\tilde{\sigma} = \frac{\sigma_0}{1 - i\omega\tau} \tag{1.14}$$

in the high-frequency limit, i.e., $\omega\tau \gg 1$, to show that $\tilde{N}$ is purely imaginary when $\omega < \omega_p$, where $\omega_p = (n_e e^2 / \varepsilon_r \varepsilon_0 m_e)^{1/2}$, $\sigma_0 = n_e e^2 \tau / m_e$ is the DC Drude conductivity, n_e is the electron density, τ is the scattering time, and m_e is the electron mass.

(d) Show that the reflectivity is unity when $\tilde{N}$ is purely imaginary.

Exercise 1.9 (Generalized momentum and the Hamiltonian)

The relativistic motion of a particle of rest mass m_0 and position $\mathbf{r} = (x, y, z)$ can be obtained from the Lagrangian

$$\mathcal{L} = -m_0 c^2 \sqrt{1 - \frac{v^2}{c^2}} \tag{1.15}$$

where c is the speed of light and $v = |\mathbf{v}| = \sqrt{\dot{x}^2 + \dot{y}^2 + \dot{z}^2}$. Find the generalized momentum p and show that the relativistic Hamiltonian becomes

$$\mathcal{H} = \frac{m_0 c^2}{\sqrt{1 - v^2/c^2}} = \sqrt{m_0^2 c^4 + p^2 c^2} \tag{1.16}$$

2

Electron Waves and Schrödinger's Equation

Learning objectives:

- Developing an understanding of when and how classical particles start behaving as quantum mechanical waves.

- Deriving the most fundamental equation in quantum mechanics, the Schrödinger equation, which determines the states and dynamics of quantum particles.

- Understanding the probabilistic meaning of the wavefunction, ψ, and learn how to calculate the expectation values of observable quantities.

- Solving the Schrödinger equation for example problems involving electrons in simple potential energy landscapes.

QUANTUM MECHANICS IS currently the most fundamental theory in use in many disciplines of science and engineering. It is particularly important when one is dealing with nanoscale and atomic-scale systems. However, many phenomena and properties that occur at atomic scales are strange and nonintuitive. There are a number of concepts that simply do not exist in the macroscopic world where we live. Wave–particle duality is one of them. In this chapter, we examine how and when classical particles start behaving as quantum mechanical waves, derive the most important wave equation that quantum particles obey, Schrödinger's equation, and solve it for the elementary problems of electron waves in given potential energy landscapes. We will also learn how to calculate the expectation values of observables when the wavefunction is known. Schrödinger's equation will be extensively used throughout the rest of this textbook. More complicated potential energy problems, particularly those relevant to materials and devices, will be dealt with in Chapters 5 and 7, building upon the formulations developed in this chapter.

2.1 Wave Equation and Wavefunction

One of the striking consequences of quantum theory is that *light behaves as particles* under certain circumstances and *electrons behave as waves* under certain circumstances. As in classical wave theory,[1] in order to describe the behaviors of any wave, we need a wave equation. Hence, in this section, after describing the concept of wave–particle duality, we will derive Schrödinger's wave equation in order to describe the wave properties of electrons in a quantitative manner.

[1] Appendix B.

2.1.1   Wave–Particle Duality

Let's first consider light. We are familiar with wave phenomena exhibited by light, including refraction, interference, and diffraction.[2] These phenomena are well explained by the theory of electromagnetism,[3] developed in the nineteenth century by people including Michael Faraday and James Maxwell. In 1900, however, a radical idea was proposed by Max Planck, when he derived the Planck distribution formula[4] to explain the colors of hot bodies. In deriving this formula, he assumed that energy is radiated in discrete quantized amounts or packets, rather than in a continuous wave.

The idea of quantization was so radical that many scientists refused to accept it, but Albert Einstein was one of the first who embraced this concept. He used it to explain other puzzling phenomena and further to make predictions that were later confirmed experimentally. In particular, Einstein explained the photoelectric effect[5] by postulating that light, or more generally all electromagnetic radiation, can be divided into a finite number of energy quanta,[6] which are now called photons. Each photon has an energy

$$\boxed{E = h\nu}, \tag{2.1}$$

where $h = 6.62607015 \times 10^{-34} \text{ m}^2 \text{ kg s}^{-1}$ is the Planck constant and ν is the frequency of light. Alternatively, using the angular frequency, $\omega = 2\pi\nu$, we can write this formula as $E = \hbar\omega$, where we introduced the reduced Planck constant, $\hbar = h/2\pi = 1.054571817 \times 10^{-34} \text{ m}^2 \text{ kg s}^{-1}$. Also, using the wavelength $\lambda = c/\nu$, where $c = 2.99792458 \times 10^8 \text{ m s}^{-1}$ is the speed of light, the photon energy is written as $E = hc/\lambda$.

[2] See Appendix B.

[3] See Appendix C.

[4] The spectral radiance

$$B(\nu, T) = \frac{2h\nu^3}{c^2} \frac{1}{e^{h\nu/k_{\mathrm{B}}T} - 1}$$

represents the spectral emissive power per unit area per unit solid angle at frequency ν emitted from a body at temperature T, where k_{B} is the Boltzmann constant, and h is the new constant Planck introduced, now called the Planck constant.

[5] A. Einstein, *Annalen der Physik* **17**, 133 (1905).

[6] This idea was even more revolutionary, and it took nearly 20 years before it was widely accepted. See G. N. Lewis, *Nature* **118**, 874 (1926), where the term *photon* was first used.

Example 2.1 Human eyes are extremely sensitive. They are known to be sensitive enough to detect as few as 100 photons per second in the middle of the visible spectral range, e.g., at a wavelength of 550 nm. This amount of light corresponds to a power of $100 \times hc/\lambda = 100 \times 6.63 \times 10^{-34} \times 2.99 \times 10^8/(550 \times 10^{-9}) = 3.6 \times 10^{-17}$ W $[= \text{J s}^{-1}]$.

Example 2.2 The solar energy at the top of the atmosphere ("AM0") is 1367 W m^{-2}, which is known as the solar constant. This number decreases to 963 W m^{-2} after the radiation has passed through the atmosphere ("AM1.5"), which corresponds to 3×10^{24} J/year if we integrate it over the entire surface of the earth. If we use 550 nm as the average of all the wavelengths contained in the solar spectrum, this corresponds to $3 \times 10^{24} \times 550 \times 10^{-9}/(6.63 \times 10^{-34} \times 2.99 \times 10^8) = 8.32 \times 10^{42}$ photons per year or 2.64×10^{35} photons per second.

On the other hand, according to the theory of special relativity, which Einstein also developed in 1905,[7] a particle of rest mass m_0 moving at speed $v = |v|$ possesses the following energy versus momentum relationship:[8]

$$E = \sqrt{c^2 p^2 + m_0^2 c^4},\tag{2.2}$$

where $p = |\boldsymbol{p}| = \gamma m_0 v$ is the relativistic momentum.[9]

By applying Equation (2.2) to photons, for which $m_0 = 0$,[10] we obtain the photon E–p relationship:

$$\boxed{E = cp}.\tag{2.3}$$

Hence, we ended up having two separately derived expressions for the photon energy, i.e., Equations (2.1) and (2.3). By equating them, we get $h\nu = cp$. Furthermore, since $c = \nu\lambda$,[11] we get

$$\boxed{\lambda = \frac{h}{p}}.\tag{2.4}$$

This result, known as the de Broglie relation, is the central equation representing the concept of wave–particle duality, symbolically connecting a wave quantity (λ)[12] and a particle quantity (p).[13]

Equation (2.4) was derived by the French physicist Louis de Broglie in 1924 in his theory of *matter* waves. Just as we did above, he combined special relativity and the Planck–Einstein light quantum formula to derive this famous formula. However, de Broglie hypothesized further that this relation should hold not only for photons but also for *all* matter; namely, that any particle moving with momentum p should have a quantum mechanical wavelength, h/p. A direct corollary of this statement is that electrons should exhibit wave-like properties! This extremely bold idea was indeed confirmed experimentally by Clinton Davisson and Lester Germer in the USA and by George Paget Thomson in the UK in 1927. These groups demonstrated that a beam of electrons is *diffracted* by crystals, which is the well-known phenomenon of Bragg diffraction[14] (demonstrated earlier using X-rays), strongly supporting de Broglie's hypothesis that electrons behave as waves.[15]

More recent experiments by Akira Tonomura and coworkers have demonstrated Young's double-slit *interference* of a beam of electrons; see Figure 2.1. Interference is another representative wave property, and thus, this experiment further confirmed the wave nature of electrons. In addition, this experiment also illuminated how individual electrons are seen to hit the screen as individual particles without any sign of interference when the number of electrons is small but

[7] A. Einstein, *Annalen der Physik* **17**, 891 (1905).

[8] See also Exercise 1.9.

[9] Here, $\gamma = \left(1 - v^2/c^2\right)^{-1/2}$.

[10] See Exercise 2.3.

[11] See Appendices B and C.

[12] Appendix B.

[13] Appendix A.

[14] See Section 7.2.

[15] A bullet of mass 40 g moving at $1000\ \mathrm{m\,s^{-1}}$ has a wavelength of $6.63 \times 10^{-34}/(5 \times 10^{-3} \times 10^3) = 1.33 \times 10^{-34}$ m. This is too small for its wave nature to be noticed. In fact, if the Planck constant, h, were zero, there would be no quantum effect at all. On the other hand, if h were larger, there would be radical changes in the macroscopic world; see, for example, P.-K. Yang, "How does Planck's constant influence the macroscopic worlds?," *European Journal of Physics* **37**, 055406 (2016).

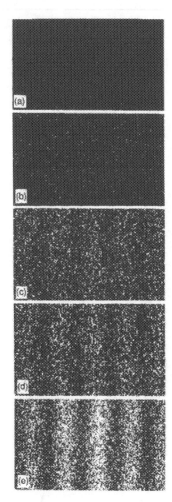

Figure 2.1 Young's double-slit experiment for an electron beam, showing the buildup of an interference pattern. The number of electrons that hit the screen was (a) 10, (b) 100, (c) 3000, (d) 20,000, and (e) 70,000. Reproduced from A. Tonomura *et al.*, *American Journal of Physics* **57**, 117 (1989), with the permission of the American Association of Physics Teachers.

[16] See Section 3.5 for further details about quantum measurement.

[17] E. Schrödinger, *Annalen der Physik* **384**, 361 (1926). For how this equation was derived, see, e.g., Arthur I. Miller, in: *It Must Be Beautiful: Great Equations of Modern Science*, edited by Graham Farmelo (Granta Books, London, 2002), pp. 110-131.

then the interference pattern emerges after a large number of electrons hit the screen. This observation on electrons is also related to Einstein's earlier statement in the context of the photoelectric effect: light intensity is proportional to the number of photons. According to the electromagnetic wave theory, the light intensity is proportional to the square of the electric field of light. According to the photon theory of light, the intensity is equal to the number of photons times the energy of each photon, which is $h\nu$.

Niels Bohr and Werner Heisenberg developed the quantum concept called complementarity. According to this foundational quantum view, the wave and particle models are complementary to each other in the following sense. If a measurement proves the wave character of radiation or matter, then it is impossible to prove the particle character in the *same* measurement. And, if a measurement reveals the particle character, the wave character cannot be seen. This view implies the important role measurement plays in atomic-scale phenomena.[16] One can design an experiment to demonstrate *either* wave properties *or* particle properties of radiation or matter, *but not both simultaneously*. The wave and particle pictures are thus mutually exclusive. However, they are not contradictory but complementary.

2.1.2 Schrödinger's Wave Equation

Stimulated by de Brogie's work on matter waves, the Austrian–Irish physicist Erwin Schrödinger developed a wave equation in 1925,[17] which has become the central equation in quantum mechanics. He started from the classical wave equation,

$$\nabla^2 \psi(\boldsymbol{r}) = -k^2 \psi(\boldsymbol{r}), \tag{2.5}$$

where ψ is a wavefunction, $k = |\boldsymbol{k}| = 2\pi/\lambda$ is the wavenumber, $\boldsymbol{k}$ is the wavevector, $\boldsymbol{r}$ is the position vector, and ∇^2 is the Laplacian. The de Broglie relation, Equation (2.4), in terms of k reads

$$p = \hbar k, \tag{2.6}$$

and thus, we can rewrite k^2 as

$$k^2 = \frac{p^2}{\hbar^2} = \frac{2m}{\hbar^2}(E - V). \tag{2.7}$$

Here, the nonrelativistic, classical-particle relation, $p^2/2m = \mathcal{T} = E - V$, for the momentum ($p$), total energy ($E$), kinetic energy ($\mathcal{T}$), and potential energy (V) has been used. If we substitute Equation (2.7) into Equation (2.5), we obtain the Schrödinger equation:

$$\boxed{-\frac{\hbar^2}{2m}\nabla^2\psi(r) + V(r)\psi(r) = E\psi(r)}.$$

(2.8)

It is important to note that the Schrödinger equation can be written as $\hat{H}\psi = E\psi$,[18] where

$$\hat{H} = -\frac{\hbar^2}{2m}\nabla^2 + V(r)$$

(2.9)

is known as the Hamiltonian operator. For a given potential energy V, and thus $\hat{H}$, one can solve the Schrödinger equation to obtain a set of eigensolutions $\{\psi_n\}_{n=1,2,\dots}$ (or "eigenfunctions") and the corresponding eigenvalues $\{E_n\}_{n=1,2,\dots}$ (or "eigenenergies").[19]

2.1.3 Wavefunction

The meaning of the wavefunction, $\psi(r)$, was initially a subject of much debate.[20] Schrödinger was trying to develop a theory that explains atomic- and subatomic-scale phenomena entirely using the classical theory of waves. As such, he was seeking an interpretation of the wavefunction purely from a wave point of view, without success. A correct interpretation of $\psi(r)$, embracing the concept of wave–particle duality, was proposed by Max Born: the square of the amplitude of the wavefunction in some specific region of space, $|\psi(r)|^2$, is related to the probability of finding the associated quantum particle in that region of space.

Further allowing the wavefunction to depend on time, t, one can state that the square of the absolute value of the wavefunction, i.e., $|\psi(x,y,z,t)|^2$ is related to the probability of finding the particle at position $r = (x,y,z)$ at time t. In order for this probabilistic interpretation to be quantitatively meaningful, the wavefunction has to be properly normalized in the following sense:

$$\int \psi(r,t)\psi^*(r,t)d^3r = \int_{-\infty}^{\infty}\int_{-\infty}^{\infty}\int_{-\infty}^{\infty}|\psi(x,y,z,t)|^2 dxdydz = 1. \quad (2.10)$$

In other words, if one integrates the square of the absolute value of the wavefunction over the entire space, it has to be equal to 1 at any moment of time t.

[18] According to linear algebra, an equation of this form is known as an eigenvalue equation. See Chapter 3 for more details.

[19] Schrödinger solved this equation for a particular potential energy, $V(r) = -e^2/(4\pi\varepsilon_0 r)$, i.e., an attractive Coulomb potential, to obtain eigenfunctions and eigenenergies that successfully explained the emission spectrum of the hydrogen atom.

[20] This was just one of the many subjects of much debate among the founders of quantum mechanics. For more details about the early disputes about quantum theory (especially those between Bohr and Einstein), see, e.g., Chapter IV of A. Pais, *"Subtle is the Lord..."*: *The Science and the Life of Albert Einstein* (Oxford University Press, 1982); see also J. Baggott, *The Meaning of Quantum Theory* (Oxford University Press, 1992), Chapter 3.

2.2   Quantum Confinement

[21] See Chapters 5 and 7.

[22] This is done by appropriately designing the potential energy term, $V(r)$, in the Hamiltonian.

[23] We will also see how different system dimensions lead to the concepts of quantum wells, quantum wires, and quantum dots, which are at the core of today's optoelectronic devices.

[24] More generally, $V = $ constant, since the choice of the zero of potential energy is arbitrary.

One important goal of the quantum engineering of solid-state devices[21] is to confine electrons in appropriate small regions within the device by creating an artificial potential energy landscape.[22] Below, let us use simple example potentials to understand how small the system size has to be for quantum confinement to occur.[23]

We begin with a completely free particle, i.e., $V = 0$ everywhere.[24] Then, the Schrödinger equation is written as

$$-\frac{\hbar^2}{2m}\nabla^2\psi(r) = E\psi(r). \tag{2.11}$$

The general solution to this differential equation is

$$\psi_{\text{free}}(r) = Ae^{ik\cdot r} + Be^{-ik\cdot r}, \tag{2.12}$$

where A and B are constants and $|k|^2 = k^2 = 2mE/\hbar^2$.[25,26]

[25] Note that the free-particle wavefunction is *not* normalizable but is useful for describing particle motion. See Section 5.1 for more about such unbound states.

[26] The two terms in Equation (2.12) represent plane waves with wavevectors $\pm k$ in the wave picture and freely moving particles with momentum $p = \hbar k$ in the particle picture.

Next, we consider the case of a particle confined in a potential box. Let us assume that the potential energy can be separated into three separate spatial directions as

$$V(r) = V(x,y,z) = V_x(x) + V_y(y) + V_z(z). \tag{2.13}$$

We assume the wavefunction to be in the form

$$\psi(r) = X(x)Y(y)Z(z) \tag{2.14}$$

and substitute it into the Schrödinger equation. Then, we get

$$-\frac{\hbar^2}{2m}\left(YZ\frac{d^2X}{dx^2} + XZ\frac{d^2Y}{dy^2} + XY\frac{d^2Z}{dz^2}\right)$$
$$+ (V_x + V_y + V_z)XYZ = EXYZ, \tag{2.15}$$

or by dividing both sides by XYZ, we get

$$\left(-\frac{\hbar^2}{2m}\frac{1}{X}\frac{d^2}{dx^2} + V_x\right)X + \left(-\frac{\hbar^2}{2m}\frac{1}{Y}\frac{d^2}{dy^2} + V_y\right)Y$$
$$+ \left(-\frac{\hbar^2}{2m}\frac{1}{Z}\frac{d^2}{dz^2} + V_z\right)Z = E, \tag{2.16}$$

which has to be satisfied for any values of x, y, and z. This is possible only when the three terms on the left hand side are constants:

$$\left(-\frac{\hbar^2}{2m}\frac{1}{X}\frac{d^2}{dx^2}+V_x\right)X=E_x,\qquad(2.17)$$

$$\left(-\frac{\hbar^2}{2m}\frac{1}{Y}\frac{d^2}{dy^2}+V_y\right)Y=E_y,\qquad(2.18)$$

$$\left(-\frac{\hbar^2}{2m}\frac{1}{Z}\frac{d^2}{dz^2}+V_z\right)Z=E_z,\qquad(2.19)$$

$$E_x+E_y+E_z=E.\qquad(2.20)$$

Let us now consider a 1D potential well with width L and an infinite potential barrier height:

$$V(x)=\begin{cases}0, & |x|\le L/2,\\ \infty, & |x|>L/2.\end{cases}\qquad(2.21)$$

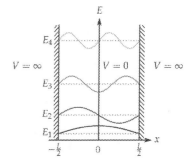

Figure 2.2 Quantum confinement in a 1D square potential well with an infinite barrier height. © Deyin Kong (Rice University).

Inside the well ($|x|\le L/2$), the potential energy is zero, so the Schrödinger equation reads

$$-\frac{\hbar^2}{2m}\frac{d^2\psi(x)}{dx^2}=E\psi(x),\qquad(2.22)$$

whose general solution is

$$\psi(x)=Ae^{ikx}+Be^{-ikx},\qquad(2.23)$$

where $k=\sqrt{2mE}/\hbar$. Since the barrier height is infinite, there is zero probability that the particle can exist outside the well. Therefore, we impose the boundary conditions that the wavefunction vanish at the edges of the well: $\psi(\pm L/2)=0$. Namely,

$$Ae^{ikL/2}+Be^{-ikL/2}=0,\qquad(2.24)$$
$$Ae^{-ikL/2}+Be^{ikL/2}=0\qquad(2.25)$$

or equivalently,

$$Ae^{ikL}+B=0,\qquad(2.26)$$
$$Ae^{-ikL}+B=0.\qquad(2.27)$$

By subtracting one equation from the other, we obtain

$$A(e^{ikL}-e^{-ikL})=2iA\sin(kL)=0.\qquad(2.28)$$

Therefore, $kL=n\pi$, where $n=1,2,3,\ldots$, and

$$\boxed{E_n=\frac{\hbar^2k^2}{2m}=\frac{\pi^2\hbar^2}{2mL^2}n^2}.\qquad(2.29)$$

The prefactor $\pi^2\hbar^2/2mL^2$ provides the characteristic quantum energy scale in the system; it determines whether quantization is apparent for a given box at a given temperature T. Quantum effects become important when this energy is comparable to or larger than the thermal energy, k_BT, where $k_B = 1.38064852 \times 10^{-23}$ m^{2}kg s^{-2}K^{-1} is the Boltzmann constant:

$$\frac{\pi^2\hbar^2}{2mL^2} \gtrsim k_BT. \qquad (2.30)$$

By rewriting this inequality, we can derive the same condition in terms of a length scale:

$$L \lesssim \frac{h}{\sqrt{8mk_BT}}. \qquad (2.31)$$

The right-hand side of the inequality (2.31) sets the critical length scale for the observability of quantum confinement effects. Namely, we can conclude that quantum effects become important when the characteristic system dimension, L, becomes comparable to or smaller than the thermal de Broglie wavelength, λ_D.[27]

In a more general 3D box with dimensions $L_x \times L_y \times L_z$, the energy depends on three indices:

$$E_{n_x,n_y,n_z} = \frac{\pi^2\hbar^2}{2mL_x^2}\, n_x^2 + \frac{\pi^2\hbar^2}{2mL_y^2}\, n_y^2 + \frac{\pi^2\hbar^2}{2mL_z^2}\, n_z^2, \qquad (2.32)$$

where $n_x, n_y, n_z = 1, 2, 3, \ldots$

Furthermore, because of the separability of motion into three spatial dimensions (x, y, and z), one can talk about quantum effects in different directions separately. If one of the spatial dimensions of the system, say, L_z, is smaller than λ_D while the other two (L_x and L_y) are much larger than λ_D, then the particle in the box is 2D in nature. Namely, the particle is quantum mechanically confined in the z-direction whereas its motion in the x–y plane is classical. As a result, $E_z = \hbar^2 k_{n_z}^2/2m$ is discrete ($n_z = 1, 2, 3, \ldots$), but $E_x = \hbar^2 k_x^2/2m$ and $E_y = \hbar^2 k_y^2/2m$ are continuous:

$$E_{n_z}(k_x, k_y) = \frac{\hbar^2}{2m}\left(k_x^2 + k_y^2\right) + \frac{\pi^2\hbar^2}{2mL_z^2}\, n_z^2: \quad \text{quantum well} \qquad (2.33)$$

Similarly, the particle motion is 1D in a quantum wire ($L_y, L_z < \lambda_D$, $L_x \gg \lambda_D$) and zero-dimensional (0D) in a quantum box/dot

[27] The thermal de Broglie wavelength, $\lambda_D := h/\sqrt{2\pi mk_BT}$, is the de Broglie wavelength of a free particle of mass m at temperature T with energy–momentum relation $E = p^2/2m$, derivable from the partition function of a 3D ideal gas. More generally, for a particle with an energy–momentum relation $E = ap^s$ in n dimensions,

$$\lambda_D = \frac{h}{\sqrt{\pi}}\left(\frac{a}{k_BT}\right)^{1/s}\left[\frac{\Gamma(\frac{n}{2}+1)}{\Gamma(\frac{n}{s}+1)}\right]^{1/n},$$

where $\Gamma(x)$ is the gamma function. See Z. Yan, *European Journal of Physics* **21**, 625 (2000).

$(L_x, L_y, L_z < \lambda_D)$:

$$E_{n_y,n_z}(k_x) = \frac{\hbar^2}{2m}k_x^2 + \frac{\pi^2\hbar^2}{2mL_y^2}\,n_y^2 + \frac{\pi^2\hbar^2}{2mL_z^2}\,n_z^2 : \quad \text{quantum wire}$$

(2.34)

$$E_{n_x,n_y,n_z} = \frac{\pi^2\hbar^2}{2mL_x^2}\,n_x^2 + \frac{\pi^2\hbar^2}{2mL_y^2}\,n_y^2 + \frac{\pi^2\hbar^2}{2mL_z^2}\,n_z^2 : \quad \text{quantum box/dot}$$

(2.35)

As in atoms, a particle confined in a quantum box is fully quantized, with no continuous variable in the specification of its energy.

2.3 Bound States

There are two types of quantum states – bound states and unbound states. The wavefunctions of bound states have zero amplitude at $|r| = \infty$ and a large amplitude in the vicinity of a potential minimum. For example, hydrogen atomic states, such as the $1s$ and $2p$ states, are bound states, their wavefunctions having large amplitudes near the proton. The wavefunctions of unbound states are extended in space but have a well-defined momentum (or wavenumber).[28] In this section, we will survey some of the basic characteristics of bound states, using the eigenfunctions and eigenenergies of a particle in a quantum well as an example.

2.3.1 Eigenfunctions

For the ground state ($n = 1$) of an infinite-barrier quantum well with width L, we have the wavenumber $k_1 = \pi/L$. Therefore, from Equations (2.26) or (2.27), we can deduce that $A = B$, and the wavefunction is written as

$$\psi_1(x) = A(e^{ik_1 x} + e^{-ik_1 x}) = 2A\cos\left(\frac{\pi}{L}x\right).$$ (2.36)

For the second state ($n = 2$), $k_2 = 2\pi/L$, so $A = -B$ results from Equation (2.26) or (2.27), and the wavefunction is

$$\psi_2(x) = A(e^{ik_2 x} - e^{-ik_2 x}) = 2iA\sin\left(\frac{2\pi}{L}x\right).$$ (2.37)

In a similar manner, we can show that $\psi_n(x)$ is proportional to either $\sin(k_n x)$ (an odd function) or $\cos(k_n x)$ (an even function). More generally, whenever the potential energy possesses inversion symmetry, i.e., $V(x) = V(-x)$ in 1D, the wavefunction must possess definite parity, i.e., either odd or even parity.[29]

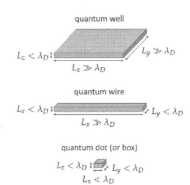

Figure 2.3 Quantum confinement in different dimensions. Quantum effects become important when the characteristic system dimension, L, is comparable with or smaller than the thermal de Broglie wavelength, λ_D. © Deyin Kong (Rice University).

[28] They are useful for describing propagating waves, such as Bloch electrons in solids, including artificial semiconductor structures such as superlattices and quantum cascade lasers (see Chapters 5 and 7).

[29] The parity operator, $\hat{\Pi}$, is defined through $\hat{\Pi}\psi(x) = \psi(-x)$. For a wavefunction with an odd (even) parity, $\hat{\Pi}\psi(x) = -\psi(x)$ ($\hat{\Pi}\psi(x) = \psi(x)$). See Appendix D.

We can intentionally break this requirement by considering the same problem but with the quantum well potential shifted in the positive x-direction by $L/2$, shown in Figure 2.4:

$$V(x) = \begin{cases} 0, & 0 \le x \le L, \\ \infty, & x < 0, x > L. \end{cases} \tag{2.38}$$

In this case, the boundary conditions are $\psi(0) = \psi(L) = 0$. Thus,

$$A + B = 0, \tag{2.39}$$

$$Ae^{ikL} + Be^{-ikL} = 0 \tag{2.40}$$

Combining these two equations, we obtain the eigenfunctions

$$\psi_n(x) = A' \sin(k_n x), \tag{2.41}$$

where $A' = 2iA$ is a constant, $k_n = n\pi/L$, and the corresponding energies (eigenenergies) are $E_n = \hbar^2 k_n^2/2m$ as before. Therefore, in this case, all wavefunctions ψ_n are proportional to $\sin(k_n x)$. We will use this type of wavefunction in the following.

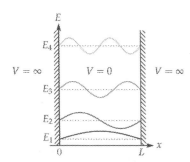

Figure 2.4 1D square potential well with an infinite barrier height with edges at $x = 0$ and L. Note that $V(x) = V(-x)$ is not satisfied in this case, so we do not have odd and even wavefunctions in contrast to the case depicted in Figure 2.2. © Deyin Kong (Rice University).

2.3.2   Orthonormality

The wavefunctions $\psi_n(x)$ have to be normalized in such a way that they satisfy Equation (2.10). Hence,

$$\int_0^L |\psi_n(x)|^2 dx = |A'|^2 \int_0^L \sin^2\left(\frac{n\pi}{L}x\right) dx$$
$$= \frac{|A'|^2}{2} \int_0^L \left\{1 - \cos\left(\frac{2n\pi}{L}x\right)\right\} dx = \frac{|A'|^2 L}{2} = 1. \tag{2.42}$$

Therefore, $A' = \sqrt{2/L}$, i.e., the following expression provides properly normalized eigenfunctions:[30]

$$\psi_n(x) = \sqrt{\frac{2}{L}} \sin\left(\frac{n\pi}{L}x\right), \tag{2.43}$$

[30] In general, we can retain a phase factor in A' as $A' = e^{i\theta}\sqrt{2/L}$. However, such a phase, θ, does not influence our discussion below.

where $n = 1, 2, 3, \ldots$ Furthermore, by using Equation (2.43), one can show the following orthogonality relation between different eigenfunctions:

$$\int_0^L \psi_n(x)\psi_m^*(x)dx = 0 \tag{2.44}$$

[31] See Example 3.4

for $n \ne m$.[31] Combining the normality and orthogonality relations, we write the orthonormality relation

$$\boxed{\int_0^L \psi_n(x)\psi_m^*(x)dx = \delta_{nm} = \begin{cases} 1, & n = m, \\ 0, & n \ne m. \end{cases}} \tag{2.45}$$

Here, δ_{nm} is Kronecker's delta.

2.3.3 Eigenenergies

Irrespective of what coordinate system we use, the energy of the n-th state in the infinite-barrier quantum well with width L is given by Equation (2.29). It is instructive to see that solving the quantum well problem is essentially finding "eigenmodes of oscillation" of an electron wave inside the well, in analogy to a string with both ends fixed. In fact, by fitting an integral number of half de Broglie wavelengths into the width L of the well, we can derive the quantization law:

$$L = n\frac{\lambda}{2} = n\frac{h}{2p} = n\frac{\pi}{k} \quad \rightarrow \quad k_n = \frac{n\pi}{L}. \tag{2.46}$$

The fact that the ground state ($n = 1$) has an energy (E_1) that is not zero deserves some consideration. Since the particle is bound by the potential, we know its x coordinate to within an uncertainty of about $\Delta x \sim L$. This rather precise knowledge of position introduces, through the uncertainty principle, a strict lower limit on the uncertainty in the momentum, Δp, on the order of h/L.[32] This in turn means that the particle cannot have zero total energy since that would mean that the uncertainty in the momentum would be zero. This is the reason why the ground state has a nonzero energy, known as the zero-point energy. In this particular case of an infinite-barrier quantum well, this energy can be calculated as

$$\Delta E \sim \frac{\Delta p^2}{2m} \sim \frac{h^2}{2mL^2}, \tag{2.47}$$

which agrees with E_1 calculated from Equation (2.29).

2.3.4 Finite-Barrier Quantum Well

Let us now consider a quantum well with a finite barrier height:

$$V(x) = \begin{cases} 0, & |x| \leq L/2, \\ V_0, & |x| > L/2, \end{cases} \tag{2.48}$$

where V_0 (> 0) is a constant. See Figure 2.5. The general solution to the corresponding Schrödinger equation can be written as

$$\psi(x) = \begin{cases} Ae^{ikx} + Be^{-ikx}, & |x| \leq L/2, \\ Ce^{\kappa x}, & x < -L/2, \\ De^{-\kappa x}, & x > L/2. \end{cases} \tag{2.49}$$

where $k = \sqrt{2mE}/\hbar$, as before, and $\kappa := \sqrt{2m(V_0 - E)}/\hbar$.

[32] Introduced by Werner Heisenberg in 1927, stating that $\Delta x \, \Delta p \gtrsim h$. See Section 3.5.2 for mathematical details using operator algebra.

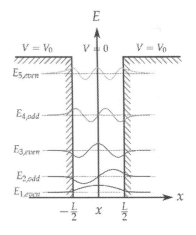

Figure 2.5 1D square potential well with a finite barrier height, V_0, with edges at $x = \pm L/2$. © Deyin Kong (Rice University).

From the symmetry of the potential profile, that is, $V(x) = V(-x)$, and parity considerations, each eigenfunction must be either an even or odd function.[33] For an even solution, we have $A\cos(kx)$ (for $|x| \leq L/2$) with $C = D$ (for $|x| > L/2$), whereas, for an odd solution, we have $A\sin(kx)$ (for $|x| \leq L/2$) with $C = -D$ (for $|x| > L/2$).

[33] See Appendix D.

Since $V(x)$ and $d^2\psi/dx^2$ are finite everywhere, ψ and $d\psi/dx$ must be continuous everywhere. Thus, the boundary conditions at $x = L/2$ for the even solutions are

$$\psi(x) = \begin{cases} \psi(L/2) = A\cos(kL/2) = Ce^{-\kappa L/2}, \\ \psi'(L/2) = -kA\sin(kL/2) = -\kappa Ce^{\kappa L/2}. \end{cases} \tag{2.50}$$

By dividing one of the equations by the other and multiplying both sides by L, we obtain

$$\xi\tan\xi = \eta, \tag{2.51}$$

where $\xi := kL/2$ and $\eta := \kappa L/2$ are dimensionless parameters. Similarly, for the odd solutions, we have

$$\xi\cot\xi = -\eta. \tag{2.52}$$

Furthermore,

$$\xi^2 + \eta^2 = \frac{mV_0L^2}{2\hbar^2}. \tag{2.53}$$

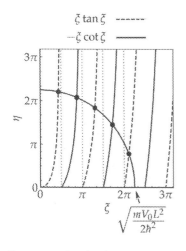

Figure 2.6 Graphical representation of Equations (2.51), (2.52), and (2.53). There exist $s+1$ bound states when $s\pi/2 < \sqrt{mV_0L^2/2\hbar^2} < (s+1)\pi/2$. © Deyin Kong (Rice University).

Equations (2.51), (2.52), and (2.53) can be solved graphically, as shown in Figure 2.6.

2.4 Expectation Values

According to the probabilistic interpretation of the wavefunction $\psi(x)$,

$$|\psi(x)|^2 = P(x) \tag{2.54}$$

is the probability *density* in the sense that the probability of finding the particle at a position between x and $x + dx$ is given by $|\psi(x)|^2dx$. Therefore, the wavefunction has to be normalized, so that

$$\int_{-\infty}^{\infty} |\psi(x)|^2 dx = 1. \tag{2.55}$$

Similarly, in 3D systems, the probability of finding the particle in an infinitesimally small region, $dxdydz$, is given by $|\psi(x,y,z)|^2dxdydz$, and the normalization requirement is

$$\iiint_{-\infty}^{\infty} |\psi(x,y,z)|^2 dxdydz = 1. \tag{2.56}$$

In the following we will discuss only the 1D case, without loss of generality. Given the above interpretation of the wavefunction, $\psi(x)$, we can calculate the most likely position as a weighted average. This is called the expectation value of x and given by

$$\langle x \rangle = \int_{-\infty}^{\infty} x P(x) dx = \int_{-\infty}^{\infty} \psi^*(x) x \psi(x) dx. \qquad (2.57)$$

So far, we have been using only the position x to evaluate the wavefunction. As we know from classical mechanics, the momentum p_x (= $\hbar k_x$) and x are canonically conjugate.[34] The wavefunction can indeed be described in momentum space (or k-space), as opposed to real (x-) space. The wavefunction evaluated in k-space, $\psi(k_x)$, has the same probabilistic interpretation as $\psi(x)$, but it is now expressed in terms of momentum. Namely, the probability that the particle has a momentum value between $\hbar k_x$ and $\hbar(k_x + dk_x)$ is given by $|\psi(k_x)|^2 dk_x$. Accordingly, the wavefunction has to be normalized, so that

$$\int_{-\infty}^{\infty} |\psi(k_x)|^2 dk_x = 1. \qquad (2.58)$$

Given the wavefunction, $\psi(k_x)$, the expectation value of momentum p_x is then calculated as

$$\langle p_x \rangle = \hbar \langle k_x \rangle = \hbar \int_{-\infty}^{\infty} \psi^*(k_x) k_x \psi(k_x) dk_x. \qquad (2.59)$$

The wavefuctions $\psi(x)$ and $\psi(k_x)$ are connected through Fourier transformation:

$$\psi(k_x) = \frac{1}{\sqrt{2\pi}} \int_{-\infty}^{\infty} \psi(x) e^{-ik_x x} dx, \qquad (2.60)$$

$$\psi(x) = \frac{1}{\sqrt{2\pi}} \int_{-\infty}^{\infty} \psi(k_x) e^{ik_x x} dk_x. \qquad (2.61)$$

By taking the complex conjugate of Equation (2.60), we obtain

$$\psi^*(k_x) = \frac{1}{\sqrt{2\pi}} \int_{-\infty}^{\infty} \psi^*(x) e^{ik_x x} dx. \qquad (2.62)$$

Also, by performing the integral on the right-hand side of Equation (2.60) by parts, we obtain

$$\psi(k_x) = -\frac{i}{\sqrt{2\pi}} \frac{1}{k_x} \int_{-\infty}^{\infty} \frac{\partial \psi(x)}{\partial x} e^{-ik_x x} dx. \qquad (2.63)$$

Further, by substituting Equations (2.62) and (2.63) into Equation (2.59),

[34] See Appendix A. In Hamilton's formulation of classical mechanics, a physical system (in 3D) is described by canonical coordinates $(q, p) = (q_1, q_2, q_3; p_1, p_2, p_3)$. Here, the q_i are called generalized coordinates (or positions), and $p_i = \partial \mathcal{L} / \partial \dot{q}_i$ are their conjugate momenta, where $\mathcal{L} = T - V$ is the Lagrangian. Once the Hamiltonian of the system, $\mathcal{H}(q, p, t) = \sum_i \dot{q}_i p_i - \mathcal{L}$, is known, the time evolution of the system is described by Hamilton's equations, i.e., Equations (A.33):

$$\dot{p} = -\frac{\partial \mathcal{H}}{\partial q},$$

$$\dot{q} = \frac{\partial \mathcal{H}}{\partial p}.$$

The coordinates q_i and p_i satisfy canonical relations:

$$\{q_i, q_j\}_{\text{PB}} = 0,$$
$$\{p_i, p_j\}_{\text{PB}} = 0,$$
$$\{q_i, p_j\}_{\text{PB}} = \delta_{ij},$$

where

$$\{u, v\}_{\text{PB}} := \sum_{i=1}^{3} \left(\frac{\partial u}{\partial q_i} \frac{\partial v}{\partial p_i} - \frac{\partial u}{\partial p_i} \frac{\partial v}{\partial q_i} \right)$$

is a Poisson bracket; see also Equation (A.34). Hamilton's/Poisson's formulation provides the most natural path from classical mechanics to quantum mechanics, as first demonstrated by Dirac.

we can rewrite the expectation value of momentum as

$$\langle p_x \rangle = -i\hbar \int_{-\infty}^{\infty} dx' \int_{-\infty}^{\infty} dx\, \psi^*(x') \frac{\partial \psi(x)}{\partial x} \frac{1}{2\pi} \int_{-\infty}^{\infty} dk_x e^{-ik_x(x-x')}$$

$$= \int_{-\infty}^{\infty} \psi^*(x) \left[-i\hbar \frac{\partial}{\partial x} \right] \psi(x) dx$$

$$= \int_{-\infty}^{\infty} \psi^*(x) \hat{p}_x \psi(x) dx, \tag{2.64}$$

where we have introduced the momentum operator

$$\boxed{\hat{p}_x := -i\hbar \frac{\partial}{\partial x}} \tag{2.65}$$

and used the relation

$$\delta(x - x') = \frac{1}{2\pi} \int_{-\infty}^{\infty} dk_x e^{-ik_x(x-x')}, \tag{2.66}$$

where $\delta(x)$ is the Dirac delta function.[35]

Equation (2.64) can be extended to calculate the expectation value of any observable, A:

$$\langle A \rangle = \int_{-\infty}^{\infty} \psi^*(x) \hat{A} \psi(x) dx \tag{2.67}$$

where $\hat{A}$ is the operator associated with A.[36] For example, for the kinetic energy, $\mathcal{T}$, the corresponding operator is defined as

$$\hat{\mathcal{T}} := \frac{\hat{p}_x^2}{2m} = -\frac{\hbar^2}{2m} \frac{\partial^2}{\partial x^2}, \tag{2.68}$$

and the expectation value is given by

$$\langle \mathcal{T} \rangle = \int_{-\infty}^{\infty} \psi^*(x) \hat{\mathcal{T}} \psi(x) dx = \int_{-\infty}^{\infty} \psi^*(x) \left[-\frac{\hbar^2}{2m} \frac{\partial^2}{\partial x^2} \right] \psi(x) dx. \tag{2.69}$$

2.5   Time-Dependent Schrödinger Equation

Let us recall the Schrödinger equation for a free particle in a 1D system, Equation (2.11), and the corresponding wavefunction, Equation (2.12). Note that k_x is related to the energy E as

$$k_x = \frac{\sqrt{2mE}}{\hbar}. \tag{2.70}$$

From Equation (2.12), we see that

$$\hat{p}_x \psi(x) = (\hbar k_x) A e^{ik_x x} + (-\hbar k_x) B e^{-ik_x x}, \tag{2.71}$$

[35] Other useful forms and properties of the delta function include:

$$\int_{-\infty}^{\infty} \delta(\alpha x) dx = \frac{1}{|\alpha|},$$

$$\int_{-\infty}^{\infty} F(x) \delta(x - a) dx = F(a).$$

[36] Note that the position x is a number, as opposed to an operator, in real space, i.e., $\hat{x} = x$. Thus, any function of x, such as the potential energy $V(x)$, is also a number, and its expectation value can be calculated as

$$\langle V \rangle = \int_{-\infty}^{\infty} \psi^*(x) \hat{V}(\hat{x}) \psi(x) dx$$

$$= \int_{-\infty}^{\infty} \psi^*(x) V(x) \psi(x) dx.$$

which suggests that the first (second) term represents a free-particle state that is traveling with momentum $\hbar k_x$ $(-\hbar k_x)$.

In analogy to the classical theory of traveling plane waves,[37] we introduce the time dependence as

$$\Psi(x,t) = Ae^{i(k_x x - \omega t)} + Be^{-i(k_x x + \omega t)}$$
$$= (Ae^{ik_x x} + Be^{-ik_x x})e^{-i\omega t}, \qquad (2.72)$$

where the temporal angular frequency, ω, and spatial angular frequency (or wavenumber), k_x, are related through a dispersion relation, or energy–momentum relation:

$$E = \hbar\omega = \frac{\hbar^2 k_x^2}{2m}. \qquad (2.73)$$

In Equation (2.72) it is clear that the first (second) term represents a free-particle state traveling to the right (left). Furthermore, this wavefunction suggests the association of k_x and ω with the following operators:

$$\hat{k}_x := -i\frac{\partial}{\partial x} \qquad (2.74)$$

$$\hat{\omega} := i\frac{\partial}{\partial t}. \qquad (2.75)$$

Together with these operators, the dispersion relation, Equation (2.73), suggests that the time-dependent Schrödinger equation for a free particle in one dimension should be

$$i\hbar\frac{\partial}{\partial t}\Psi(x,t) = -\frac{\hbar^2}{2m}\frac{\partial^2}{\partial x^2}\Psi(x,t). \qquad (2.76)$$

In the presence of a conservative force field, $F(x,t) = -\partial V(x,t)/\partial x$, the dispersion relation is modified to

$$E = \hbar\omega = \frac{\hbar^2 k_x^2}{2m} + V, \qquad (2.77)$$

and thus, the energy operator (or the Hamiltonian) can be written as

$$\hat{H} = \hbar\hat{\omega} = \frac{\hbar^2 \hat{k}_x^2}{2m} + V. \qquad (2.78)$$

Combining Equations (2.74), (2.75), and (2.78), we obtain the full, time-dependent Schrödinger equation in one dimension:

$$\boxed{i\hbar\frac{\partial}{\partial t}\Psi(x,t) = \left\{-\frac{\hbar^2}{2m}\frac{\partial^2}{\partial x^2} + V(x,t)\right\}\Psi(x,t) = \hat{H}\Psi(x,t)}. \qquad (2.79)$$

This equation tells us that the time evolution of a quantum system is continuous and deterministic. Namely, Equation (2.79) for a given Hamiltonian, $\hat{H}(x,t)$, can be solved for $\Psi(x,t)$ as

$$\Psi(x,t) = \hat{U}(t)\Psi(x,0), \tag{2.80}$$

where the unitary operator

$$\hat{U}(t) := e^{-i\hat{H}t/\hbar} = 1 - \frac{i\hat{H}t}{\hbar} + \frac{1}{2}\left(\frac{i\hat{H}t}{\hbar}\right)^2 - \frac{1}{3!}\left(\frac{i\hat{H}t}{\hbar}\right)^3 + \cdots \tag{2.81}$$

is known as the time evolution operator.

When the potential energy happens to be stationary, i.e., $V(x,t) = V(x)$, Equation (2.79) can be separated into two equations describing the spatial and temporal parts, respectively. We set $\Psi(x,t) = \psi(x)\phi(t)$, substitute it in Equation (2.79), divide both sides by $\psi\phi$, and obtain

$$i\hbar\frac{1}{\phi(t)}\frac{d\phi(t)}{dt} = -\frac{\hbar^2}{2m}\frac{1}{\psi(x)}\frac{d^2\psi(x)}{dx^2} + V(x). \tag{2.82}$$

The left-hand side depends only on t while the right-hand side depends only on x. In order for this equation to hold for any values of t and x, both sides must be equal to a constant (E). Thus,

$$i\hbar\frac{d\phi}{dt} = E\phi, \tag{2.83}$$

$$-\frac{\hbar^2}{2m}\frac{d^2\psi}{dx^2} + V\psi = E\psi. \tag{2.84}$$

Equation (2.84) is the familiar time-independent Schrödinger equation in one dimension.

Remark 2.1 Time-dependent stationary state

Equation (2.83) yields

$$\phi(t) = Ae^{-iEt/\hbar} = Ae^{-i\omega t}. \tag{2.85}$$

This harmonically oscillating, time-dependent phase factor, $e^{-i\omega t}$, always exists for a *stationary* state. For example, the first three wavefunctions of the infinite-barrier square potential well can now be written

$$\Psi_1(x,t) = \sqrt{\frac{2}{L}}\sin\left(\frac{\pi}{L}x\right)e^{-i\omega_1 t}, \tag{2.86}$$

$$\Psi_2(x,t) = \sqrt{\frac{2}{L}}\sin\left(\frac{2\pi}{L}x\right)e^{-i\omega_2 t}, \tag{2.87}$$

$$\Psi_3(x,t) = \sqrt{\frac{2}{L}}\sin\left(\frac{3\pi}{L}x\right)e^{-i\omega_3 t}, \tag{2.88}$$

where $\omega_1 = E_1/\hbar = (\hbar^2/2m)(\pi^2/L^2)/\hbar$, $\omega_2 = 4\omega_1$, and $\omega_3 = 9\omega_1$. This time dependence is not observable when the particle is in a single state, as can be seen by the fact that for any state $|\Psi_n(x,t)|^2 = |\psi_n(x)|^2$. However, when the particle is in a *superposition state*, time dependence does appear. For example, when the particle is in

$$\Psi = \frac{1}{\sqrt{2}}(\Psi_1 + \Psi_2), \tag{2.89}$$

the probability density $|\Psi_n(x,t)|^2$ does oscillate as a function of time, with frequency given by $\omega_2 - \omega_1$:

$$|\Psi|^2 = |\psi_1|^2 + |\psi_2|^2 + 2\,\mathrm{Re}(\psi_1^*\psi_2)\cos(\omega_2 - \omega_1)t. \tag{2.90}$$

2.6 Chapter Summary

Starting from the concept of wave-particle duality, we introduced the fundamental wave equation, the Schrödinger equation. This is the most important equation in quantum mechanics, and it is not an exaggeration to say that practically all quantum mechanical problems involve solving the Schrödinger equation for a given potential energy. The Schrödinger equation is an eigenvalue equation, and solving it produces a set of eigenfunctions and corresponding eigenenergies. We solved the Schrödinger equation for representative situations, obtaining bound state wavefunctions and energies. Most importantly, for a 1D square-well potential with an infinitely high barrier (Figure 2.4), we obtained analytically the normalized eigenfunctions [Equation (2.43)] and their eigenenergies [Equation (2.29)] in a way similar to that used for finding the vibrational eigenfrequencies for a string with both ends fixed. The expression for the eigenenergies provided a clear physical condition under which quantum effects become important: when the system size becomes comparable with, or smaller than, the de Broglie wavelength of the particle. The wavefunction, $\psi(r)$, has to be normalized in such a way that the integral of $|\psi(r)|^2$ over all space is equal to 1. This ensures that the quantity $|\psi(r)|^2 d^3r$ can be interpreted as the probability of finding the particle at position r. The probabilistic interpretation of ψ also led to the idea of expectation values and corresponding operator expressions for observable quantities. The striking fact that particles exhibit wave properties, such as coherent superposition and interference phenomena, in the quantum world is at the core of the current quantum revolution, and will be explored further in the remaining chapters.

2.7 Exercises

Exercise 2.1 (Rest energy)

Show that in the nonrelativistic limit (i.e., $v \ll c$), Equation (2.2) reduces to

$$E = m_0 c^2 + \frac{1}{2} m_0 v^2 \tag{2.91}$$

Here, the first term is the rest energy (which any particle with mass, or 'massive particle', has), and the second term is the kinetic energy.

Exercise 2.2 (Relativistic total energy)

From the relativistic momentum $p = \gamma m_0 v$ and Equation (2.2), derive the following:

$$E = \frac{m_0 c^2}{\sqrt{1 - v^2/c^2}} \tag{2.92}$$

Exercise 2.3 (Compton scattering)

In 1923, by analyzing the X-ray scattering by graphite as a photon–electron collision process, Compton firmly established the fact that light behaves as particles with well-defined energy and momenta. By requiring that the energy, E, and momentum, p, must be finite for photons (for which $v = c$), show that photons are massless, i.e., $m_0 = 0$.

Figure 2.7 Compton scattering. An incident photon is scattered by an electron at rest, transferring part of its energy (E) and momentum (p) to the electron. © Deyin Kong (Rice University).

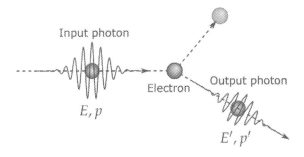

Exercise 2.4 (Matter waves)

Calculate the de Broglie wavelengths of the following particles:

(a) A bullet of mass 10 g with a speed of 100 m s^{-1}.

(b) An electron with a velocity of $3 \times 10^7 \text{ m s}^{-1}$.

(c) An electron with a kinetic energy of 15 keV.

(d) A proton with a kinetic energy of 15 eV.

Exercise 2.5 (Quantum confinement)

In GaAs, the effective mass of electrons in the conduction band, m_e^*, is $0.067m_e$, where $m_e = 9.11 \times 10^{-31}$ kg. Assuming an infinite-barrier square-quantum-well model for the potential, calculate the well size at which quantum effects become important at temperatures of 4 K, 77 K, and 300 K.

Exercise 2.6 (Band gap engineering)

The effective band gap of a GaAs/AlAs quantum well is given by

$$E_g^* = E_g + E_1^e + E_1^h,$$

where E_g is the band gap of GaAs at room temperature (1.39 eV) and E_1^e and E_1^h are the ground state energies in the conduction and valence band quantum wells. Use a simple infinite barrier approximation, and assume that the effective masses are $m_e^* = 0.067m_e$ and $m_h^* = 0.4m_e$.

(a) Calculate the quantum well width in which the effective band gap is 1.6 eV.

(b) If the size of the well width has an uncertainty comparable with a monolayer thickness (± 0.286 nm), what is the resulting uncertainty in the wavelength of band-edge emission from this quantum well?

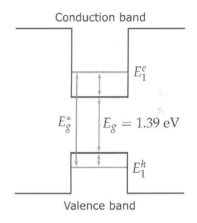

Figure 2.8 Quantum confinement increases the band gap from E_g to E_g^*. © Deyin Kong (Rice University).

Exercise 2.7 (Finite-barrier quantum well)

The following questions refer to the 1D finite-barrier square potential well depicted in Figure 2.5.

(a) When $mV_0L^2/2\hbar^2 = 16$, how many bound states exist?

(b) When $mV_0L^2/2\hbar^2 = 16$, determine the energies of the bound states. Express the energies in terms of V_0.

(c) For an electron in a GaAs quantum well with $Al_yGa_{1-y}As$ barriers, $m = 0.07m_e$, where $m_e = 9.11 \times 10^{-31}$ kg is the mass of an electron in vacuum, and $V_0 \approx y$ eV. If $L = 10$ nm, what value of y makes $mV_0L^2/2\hbar^2 = 16$?

(d) When the above-obtained value is used for y, what is the wavelength of light that will resonantly excite the electron from the lowest-energy bound state to the first excited bound state?

Exercise 2.8 (Asymmetric quantum well)

Consider a particle of mass m in a 1D potential well $V(x) = \infty$ $(x \leq 0)$, $-V_0$ $(0 < x \leq L)$, and 0 $(x > L)$, where $V_0 > 0$, depicted in Figure 2.9.

Figure 2.9 Asymmetric 1D quantum well potential. © Deyin Kong (Rice University).

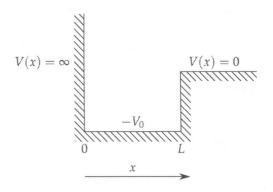

(a) Show that the bound state energies ($-V_0 < E < 0$) are given by the equation

$$\eta = -\xi \cot \xi \qquad (2.93)$$

with appropriately defined dimensionless parameters, η and ξ, that contain E.

(b) In a similar way to that used for the problem of a symmetric well with a finite barrier height, develop a graphical method to determine the bound state energies E.

(c) Calculate the minimum value of V_0 (in terms of m, L, and $\hbar$) such that the particle will have one bound state; then calculate the minimum value for two bound states. From these two results, try to obtain the lowest value of V_0 such that the system has n bound states.

Exercise 2.9 (Simple harmonic oscillator wavefunction)

Let us consider a 1D simple harmonic oscillator (SHO) consisting of a particle of mass m acted on by a restoring force of force constant κ_0, $F = -\kappa_0 x$ (Hooke's law). Since the force F is related to the potential energy V through $F = -\partial V / \partial x$, we can derive $V = \frac{1}{2}\kappa_0 x^2$. Therefore, the Hamiltonian of this system can be written as

$$\hat{H}_{\text{SHO}} = -\frac{\hbar^2}{2m}\frac{d^2}{dx^2} + \frac{1}{2}\kappa_0 x^2. \qquad (2.94)$$

The wavefunction for the lowest energy state of this system can be expressed as

$$\psi(x) = A \exp\left(-\frac{\sqrt{\kappa_0 m}}{2\hbar} x^2\right), \qquad (2.95)$$

where A is a real constant.

(a) Write down the Schrödinger equation based on Equation (2.94), and verify that Equation (2.95) is a solution.

(b) What is the eigenenergy corresponding to this wavefunction?

(c) Evaluate the probability density, $P(x) = |\psi(x)|^2$, and plot it as a function of x.

(d) Normalize the wavefunction by determining the value of A.

Exercise 2.10 (Quantum dot)

A particle is confined in a cubic quantum dot with $-L/2 < x, y, z < L/2$. The ground-state wave function is proportional to

$$\cos \frac{\pi x}{L} \cos \frac{\pi y}{L} \cos \frac{\pi z}{L}$$

(a) Find the normalized wavefunction for the ground state.

(b) What is the probability of finding this particle within the cube of $-L/4 < x, y, z < L/4$?

Exercise 2.11 (Size-dependent colors)

Describe how size-dependent coloration occurs in the following three cases: (a) CdSe quantum dots, (b) gold nanoparticles, and (c) metallic carbon nanotubes.[38]

Exercise 2.12 (Expectation values and uncertainty)

Recall that the normalized wavefunction for a particle in the nth eigenstate in a 1D square quantum well with an infinite barrier height located at $0 \leq x \leq L$ is given by Equation (2.43).

(a) When the particle is in the $n = 1$ state, calculate the following expectation values for the position x and momentum p: $\langle x \rangle$, $\langle x^2 \rangle$, $\langle p \rangle$, and $\langle p^2 \rangle$.

(b) By defining the uncertainties in the position and momentum of the particle as $\Delta x := \sqrt{\langle x^2 \rangle - \langle x \rangle^2}$ and $\Delta p := \sqrt{\langle p^2 \rangle - \langle p \rangle^2}$, respectively, calculate the product of the uncertainties $\Delta x \Delta p$.

(c) Repeat (b) and (c) when the particle is in the $n = 2$ state.

[38] Reference: E. H. Hároz *et al.*, *Journal of the American Chemical Society* **134**, 4461 (2012).

3
Mathematical Machinery and Conceptual Foundations

Learning objectives:

- Becoming proficient in operator algebra.
- Understanding the simple harmonic oscillator problem using operators.
- Developing a grasp of quantum notions such as observables, Hilbert space, superposition, and wavefunction collapse.
- Becoming fully conversant with Dirac notation.
- Learning the matrix formulation of quantum mechanics.

A REGRETTABLE AMOUNT of mathematical machinery goes into a good understanding of quantum mechanics. This could be avoided if a good intuitive understanding of many quantum systems was possible, but as intuition is generally derived from daily experience (which is governed by classical laws), we cannot expect this to be the case in general. Here, we present an in-depth introduction to the mathematical foundations of quantum mechanics, accompanied, wherever appropriate, by detailed explanations of relevant quantum concepts such as superposition, wavefunction collapse, and the uncertainty principle. As an additional benefit, the language developed in this chapter will be especially useful for describing quantum information science in Chapter 4.

3.1 Operator Algebra

Many physical quantities in quantum mechanics are expressed as operators. Throughout this book, we use a hat ˆ to indicate that the quantity is an operator. For example, $\hat{x}$, $\hat{p}$, and $\hat{L}$ are position, momentum, and angular momentum operators, respectively.

An operator changes a function into another function. In the following example, the operator $\hat{A}$ changes the function $f(x)$ into $g(x)$:

$$\hat{A}f(x) = g(x). \tag{3.1}$$

For example, if $\hat{A} = 1 + d/dx$,

$$g(x) = \left(1 + \frac{d}{dx}\right) f(x) = f(x) + \frac{df}{dx}, \tag{3.2}$$

and if $\hat{A} = (d/dx)x$,

$$g(x) = \frac{d}{dx}[xf(x)] = f(x) + x\frac{df}{dx} = \left(1 + x\frac{d}{dx}\right)f(x). \tag{3.3}$$

The second example above, Equation (3.3), indicates that we can write the following operator equation:[1]

$$\frac{d}{dx}x = 1 + x\frac{d}{dx}. \tag{3.4}$$

[1] Note that both sides of this equation are operators. In general, if an operator equation, $\hat{A} = \hat{B}$, holds, then for any function f, $\hat{A}f = \hat{B}f$.

3.1.1 Linear Operators

In quantum mechanics, practically all the operators we deal with are linear operators. If $\hat{A}$ and $\hat{B}$ are linear operators,

$$\hat{A}(cf(x)) = c\hat{A}f(x), \tag{3.5}$$

$$\hat{A}[f(x) + g(x)] = \hat{A}f(x) + \hat{A}g(x), \tag{3.6}$$

$$(\hat{A} + \hat{B})f(x) = \hat{A}f(x) + \hat{B}f(x). \tag{3.7}$$

where c is a number. The first property is called homogeneity, and the second property is called additivity.

Example 3.1 Differentiation is a linear operator, so we have

$$\frac{d}{dx}(\alpha f) = \alpha\frac{df}{dx}, \tag{3.8}$$

$$\frac{d}{dx}(f + g) = \frac{df}{dx} + \frac{dg}{dx}. \tag{3.9}$$

Example 3.2 Indefinite integration is a linear operator, so we have

$$\int \alpha f(x)\,dx = \alpha \int f(x)\,dx, \tag{3.10}$$

$$\int \{f(x) + g(x)\}\,dx = \int f(x)\,dx + \int g(x)\,dx. \tag{3.11}$$

3.1.2 Example: 1D Simple Harmonic Oscillator

To become familiar with operator algebra, here we examine the problem of the 1D simple harmonic oscillator (SHO).[2] Many of the methods and equations you will encounter in this section are useful in surprisingly diverse contexts, including photons in cavities, lattice vibrations (phonons) in crystals, an electron in a magnetic field, and collective spin wave excitations (magnons), just to name a few.

According to Hooke's law, the restoring force on a particle of mass m connected to a spring of constant κ_0, when it is displaced by x from

[2] See Exercise 2.9 for an introduction to this problem without using operator algebra.

the equilibrium position, is $F_{\text{SHO}} = -\partial V_{\text{SHO}}(x)/\partial x = -\kappa_0 x$. This suggests that the SHO potential energy $V_{\text{SHO}}(x)$ is given by $\frac{1}{2}\kappa_0 x^2$. Therefore, the 1D SHO Hamiltonian can be written as

$$\hat{H}_{\text{SHO}} = -\frac{\hbar^2}{2m}\frac{d^2}{dx^2} + \frac{1}{2}m\omega_0^2 x^2, \tag{3.12}$$

where $\omega_0 := \sqrt{\kappa_0/m}$ is the characteristic frequency of the SHO.[3] We introduce a length, ℓ_{SHO}, such that the coefficients of the two terms become equal in magnitude after a change of variables through $x = \ell_{\text{SHO}}\xi$, where ξ is a dimensionless coordinate. The Hamiltonian is written in the new variable as

$$\hat{H}_{\text{SHO}} = -\frac{\hbar^2}{2m}\frac{1}{\ell_{\text{SHO}}^2}\frac{d^2}{d\xi^2} + \frac{1}{2}m\omega_0^2\ell_{\text{SHO}}^2\xi^2, \tag{3.13}$$

and thus we choose ℓ_{SHO} such that

$$\frac{\hbar^2}{2m}\frac{1}{\ell_{\text{SHO}}^2} = \frac{1}{2}m\omega_0^2\ell_{\text{SHO}}^2 \tag{3.14}$$

or,

$$\ell_{\text{SHO}} := \sqrt{\frac{\hbar}{m\omega_0}}. \tag{3.15}$$

With this definition of ℓ_{SHO},[4] $\hat{H}_{\text{SHO}}$ is written in a compact form:

$$\hat{H}_{\text{SHO}} = \frac{\hbar\omega_0}{2}\left(-\frac{d^2}{d\xi^2} + \xi^2\right). \tag{3.17}$$

Knowing $x^2 - y^2 = (x - y)(x + y)$ from elementary algebra for ordinary numbers, x and y, we are tempted to write the following operator equation:

$$-\frac{d^2}{d\xi^2} + \xi^2 \stackrel{?}{=} \left(-\frac{d}{d\xi} + \xi\right)\left(\frac{d}{d\xi} + \xi\right). \tag{3.18}$$

However, simple operator algebra tells us that

$$\left(-\frac{d}{d\xi} + \xi\right)\left(\frac{d}{d\xi} + \xi\right)f = -\frac{d^2 f}{d\xi^2} + \xi^2 f - f, \tag{3.19}$$

$$\left(\frac{d}{d\xi} + \xi\right)\left(-\frac{d}{d\xi} + \xi\right)f = -\frac{d^2 f}{d\xi^2} + \xi^2 f + f. \tag{3.20}$$

for an arbitrary function $f(\xi)$. Thus, the Schrödinger equation for the 1D SHO can be written as

$$\hat{H}_{\text{SHO}}\psi(\xi) = \frac{\hbar\omega_0}{2}\left[\left(-\frac{d}{d\xi} + \xi\right)\left(\frac{d}{d\xi} + \xi\right) + 1\right]\psi(\xi) = E\psi(\xi). \tag{3.21}$$

[3] Unlike the square-shaped potential wells we studied in Section 2.3, what we have here is a *parabolic* potential well, proportional to x^2.

[4] The cyclotron motion of an electron in a magnetic field, B, can be described in a similar manner to SHO, and the corresponding length scale

$$\ell_B = \sqrt{\frac{\hbar}{m_e\omega_c}} = \sqrt{\frac{\hbar}{eB}} \tag{3.16}$$

is the quantum mechanical cyclotron radius, known as the magnetic length. Here, $\omega_c = eB/m_e$ is the cyclotron frequency.

We can further modify this equation into a simpler form:

$$\left(\hat{b}^\dagger \hat{b} + \frac{1}{2}\right)\psi(\xi) = \varepsilon\psi(\xi), \tag{3.22}$$

where

$$\hat{b}^\dagger := \frac{1}{\sqrt{2}}\left(-\frac{d}{d\xi} + \xi\right), \tag{3.23}$$

$$\hat{b} := \frac{1}{\sqrt{2}}\left(\frac{d}{d\xi} + \xi\right), \tag{3.24}$$

$$\varepsilon := \frac{E}{\hbar\omega_0}. \tag{3.25}$$

Furthermore, on shifting the origin of energy by setting $\varepsilon' := \varepsilon - \frac{1}{2}$, the Schrödingier equation simplifies to[5]

$$\boxed{\hat{b}^\dagger \hat{b}\,\psi = \varepsilon'\psi}. \tag{3.26}$$

[5] Note that we have not made any approximation so far, therefore Equation (3.26) is fully equivalent to the original Schrödinger equation.

Looking at Equations (3.19) and (3.20), one can see that the operators $\hat{b}^\dagger$ and $\hat{b}$ satisfy the following commutation relation:[6,7]

$$[\hat{b}, \hat{b}^\dagger] := \hat{b}\hat{b}^\dagger - \hat{b}^\dagger\hat{b} = 1. \tag{3.27}$$

By operating $\hat{b}^\dagger$ on Equation (3.26) and using Equation (3.27), we can show that

$$\hat{b}^\dagger \hat{b}\,(\hat{b}^\dagger\psi) = (\varepsilon' + 1)(\hat{b}^\dagger\psi). \tag{3.28}$$

[6] See Remark 3.1 for more details about commutation relations (or commutators).

[7] Strictly speaking, the "1" on the right-hand side here is the identity *operator* and should be written "$\hat{1}$". We will make explicit use of $\hat{1}$ only in special cases, such as in Equation (3.103), since there are no practical differences between 1 and $\hat{1}$.

This equation demonstrates that $\hat{b}^\dagger\psi$ is another eigensolution of the Schrödinger equation with an eigenenergy of $\varepsilon' + 1$. More explicitly, if ψ_n is the nth solution with energy ε'_n, $\hat{b}^\dagger$ produces the $(n+1)$th state $\psi_{n+1} = \hat{b}^\dagger\psi_n$ with energy $\varepsilon'_{n+1} = \varepsilon'_n + 1$. In this sense, $\hat{b}^\dagger$ raises the quantum index by one through the creation of one quantum ($\hbar\omega_0$). Thus, $\hat{b}^\dagger$ is called a creation operator. Similarly, one can show that

$$\hat{b}^\dagger \hat{b}\,(\hat{b}\psi) = (\varepsilon' - 1)(\hat{b}\psi). \tag{3.29}$$

In this sense, $\hat{b}$ is called an annihilation operator.

Combining these two results, we can now create a ladder of states by using these operators:

$$\cdots$$
$$\psi_{n+2} = \hat{b}^\dagger \psi_{n+1} = \hat{b}^\dagger \hat{b}^\dagger \psi_n, \quad \varepsilon'_{n+2} = \varepsilon'_{n+1} + 1 = \varepsilon'_n + 2$$
$$\psi_{n+1} = \hat{b}^\dagger \psi_n, \quad \varepsilon'_{n+1} = \varepsilon'_n + 1$$
$$\psi_n, \quad \varepsilon'_n$$
$$\psi_{n-1} = \hat{b} \psi_n, \quad \varepsilon'_{n-1} = \varepsilon'_n - 1$$
$$\psi_{n-2} = \hat{b} \psi_{n-1} = \hat{b} \hat{b} \psi_n, \quad \varepsilon'_{n-2} = \varepsilon'_{n-1} - 1 = \varepsilon'_n - 2$$
$$\cdots$$

This ladder can go up to $n = +\infty$, i.e., there is no upper limit. However, since energy has to be positive, there must be a lower limit that defines the lowest-energy state, i.e., the ground state. We define the ground state ψ_0 by requiring

$$\hat{b}\psi_0 = 0. \tag{3.30}$$

This ensures that $\varepsilon'_0 = 0$, which means that $\varepsilon_0 = \frac{1}{2}$ or $E_0 = \frac{1}{2}\hbar\omega_0$. By using Equation (3.24), we can write Equation (3.30) as

$$\frac{d}{d\xi}\psi_0 = -\xi\psi_0, \tag{3.31}$$

whose solution is

$$\psi_0(\xi) = A_0 \exp\left(-\frac{\xi^2}{2}\right), \tag{3.32}$$

where A_0 is the normalization constant. The function $\psi_0(\xi)$ is a Gaussian, centered at $\xi = 0$ (and thus $x = 0$).[8]

Once the ground state is known, we can obtain any state by successive operations of $\hat{b}^\dagger$. For example,

$$\psi_1(\xi) = \hat{b}^\dagger \psi_0 = A_1 \times 2\xi \exp\left(-\frac{\xi^2}{2}\right), \tag{3.33}$$

$$\psi_2(\xi) = \hat{b}^\dagger \psi_1 = A_2 \times (4\xi^2 - 2) \exp\left(-\frac{\xi^2}{2}\right). \tag{3.34}$$

We note that $\psi_1(\xi)$ is an odd function crossing zero at $\xi = 0$, while $\psi_2(\xi)$ is an even function with two peaks, two zeros, and one valley at $\xi = 0$. The n-th state can be written as

$$\psi_n(\xi) = (\hat{b}^\dagger)^n \psi_0 = A_n \times H_n(\xi) \exp\left(-\frac{\xi^2}{2}\right), \tag{3.35}$$

where $H_n(\xi)$ is the nth Hermite polynomial,

$$H_n(\xi) = (-1)^n e^{\xi^2/2} \frac{d^n}{d\xi^n} e^{-\xi^2/2}, \tag{3.36}$$

and the normalization constant A_n is given by

$$A_n = \frac{1}{\sqrt{2^n n!}\,\ell_{\text{SHO}}}\frac{1}{\pi^{1/4}} = \frac{1}{\sqrt{2^n n!}}\left(\frac{m\omega_0}{\pi\hbar}\right)^{1/4}. \qquad (3.37)$$

The corresponding energies are

$$\boxed{E_n = \left(n + \frac{1}{2}\right)\hbar\omega_0}, \qquad (3.38)$$

where $n = 0, 1, \ldots$ Figure 3.1 shows the eigenfunctions and eigen-energies of the lowest four states.

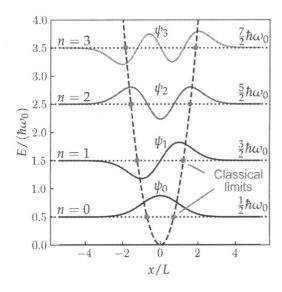

Figure 3.1 Energies and wavefunctions of eigenstates of a 1D simple harmonic oscillator. © Deyin Kong (Rice University).

Remark 3.1 Commutators

The commutation relation between two linear operators $\hat{A}$ and $\hat{B}$ informs us whether their order is changeable. Namely, when the commutator of $\hat{A}$ and $\hat{B}$, defined as

$$[\hat{A}, \hat{B}] := \hat{A}\hat{B} - \hat{B}\hat{A}, \qquad (3.39)$$

is zero, one can change their order, and these operators are then said to *commute*. On the other hand, when $[\hat{A}, \hat{B}] \neq 0$, $\hat{A}$ and $\hat{B}$ do not commute, and their order is crucial.[9]

The most important commutator is that between $\hat{x}$ (position) and $\hat{p}_x$ (linear momentum), $[\hat{x}, \hat{p}_x]$. For any wavefunction $\psi(x)$,

$$[\hat{x}, \hat{p}_x]\psi = \hat{x}\hat{p}_x\psi - \hat{p}_x\hat{x}\psi = x(-i\hbar)\frac{d\psi}{dx} - (-i\hbar)\frac{d}{dx}(x\psi)$$

$$= (-i\hbar)\left(x\frac{d\psi}{dx} - x\frac{d\psi}{dx}\right) + i\hbar\psi = i\hbar\psi. \qquad (3.40)$$

[9] As we will see in Section 3.5.2, two observables whose corresponding operators do not commute with each other must obey Heisenberg's uncertainty principle.

Hence,

$$[\hat{x}, \hat{p}_x] = i\hbar. \tag{3.41}$$

Further, by recalling $x = \ell_{\text{SHO}}\xi$ (and thus $d/d\xi = \ell_{\text{SHO}}d/dx$), where $\ell_{\text{SHO}} = \sqrt{\hbar/m\omega_0}$, we can rewrite the annihilation and creation operators for the 1D SHO, Equations (3.23) and (3.24), as

$$\hat{b} = \frac{1}{\sqrt{2}}\left(\ell_{\text{SHO}}\frac{d}{dx} + \frac{x}{\ell_{\text{SHO}}}\right) = \frac{i}{\sqrt{2m\hbar\omega_0}}(\hat{p}_x - im\omega_0\hat{x}), \tag{3.42}$$

$$\hat{b}^\dagger = \frac{1}{\sqrt{2}}\left(-\ell_{\text{SHO}}\frac{d}{dx} + \frac{x}{\ell_{\text{SHO}}}\right) = \frac{1}{\sqrt{2m\hbar\omega_0}}(\hat{p}_x + im\omega_0\hat{x}). \tag{3.43}$$

Alternatively, we can write

$$\hat{x} = i\sqrt{\frac{\hbar}{2m\omega_0}}\,(\hat{b} - \hat{b}^\dagger), \tag{3.44}$$

$$\hat{p}_x = \sqrt{\frac{m\hbar\omega_0}{2}}\,(\hat{b} + \hat{b}^\dagger), \tag{3.45}$$

with which we can obtain the following relation:

$$\begin{aligned}
[\hat{x}, \hat{p}_x] &= i\sqrt{\frac{\hbar}{2m\omega_0}}\sqrt{\frac{m\hbar\omega_0}{2}}\,[(\hat{b} - \hat{b}^\dagger), (\hat{b} + \hat{b}^\dagger)] \\
&= \frac{i\hbar}{2}\left\{(\hat{b} - \hat{b}^\dagger)(\hat{b} + \hat{b}^\dagger) - (\hat{b} + \hat{b}^\dagger)(\hat{b} - \hat{b}^\dagger)\right\} \\
&= \frac{i\hbar}{2}2(\hat{b}\hat{b}^\dagger - \hat{b}^\dagger\hat{b}) = i\hbar[\hat{b}, \hat{b}^\dagger].
\end{aligned} \tag{3.46}$$

Therefore, we see that the commutation relation between $\hat{b}$ and $\hat{b}^\dagger$ that we obtained earlier, i.e., $[\hat{b}, \hat{b}^\dagger] = 1$, Equation (3.27), is equivalent to the commutation relation between $\hat{x}$ and $\hat{p}_x$, i.e., Equation (3.41).

3.2 Hermiticity and Observables

It is straightforward to see that the creation and annihilation operators, $\hat{b}^\dagger$ and $\hat{b}$, of the SHO, defined by Equations (3.23) and (3.24), respectively, satisfy the following relation:

$$\begin{aligned}
\int_{-\infty}^{\infty}\phi^*(\xi)\,\hat{b}\,\psi(\xi)d\xi &= \int_{-\infty}^{\infty}\phi^*\frac{1}{\sqrt{2}}\left(\frac{d\psi}{d\xi} + \xi\psi\right)d\xi \\
&= \frac{1}{\sqrt{2}}\left\{[\phi^*\psi]_{-\infty}^{\infty} - \int_{-\infty}^{\infty}\frac{d\phi^*}{d\xi}\psi d\xi + \int_{-\infty}^{\infty}\xi\phi^*\psi d\xi\right\} \\
&= \frac{1}{\sqrt{2}}\int_{-\infty}^{\infty}\left(-\frac{d}{d\xi} + \xi\right)\phi^*\psi d\xi \\
&= \int_{-\infty}^{\infty}(\hat{b}^\dagger\phi)^*\psi d\xi.
\end{aligned} \tag{3.47}$$

Here, ϕ and ψ are arbitrary normalized functions, and they vanish as $x \to \pm\infty$. This relationship demonstrates that the two operators $\hat{b}^\dagger$

and $\hat{b}$ are related to each other in a special manner. In fact, it can be shown that, for any linear operator $\hat{A}$, there always exists $\hat{B}$ such that

$$\int \phi^*(\hat{A}\psi)d^3r = \int (\hat{B}\phi)^*\psi d^3r; \qquad (3.48)$$

the operator $\hat{B}$ is called the hermitian adjoint (or hermitian conjugate) of $\hat{A}$ and denoted as $\hat{A}^\dagger$.[10] If a linear operator $\hat{A}$ is its own hermitian adjoint, i.e., $\hat{A} = \hat{A}^\dagger$, then $\hat{A}$ is said to be a hermitian operator.[11] In other words, Equation (3.48) tells us that, if $\hat{A}$ is hermitian, then, for two arbitrary functions ϕ and ψ,

$$\int \phi^*(\hat{A}\psi)d^3r = \int (\hat{A}\phi)^*\psi d^3r. \qquad (3.49)$$

[10] In this sense, $\hat{b}^\dagger$ is the hermitian adjoint of $\hat{b}$. One can also see that $\hat{b}$ is the hermitian adjoint of $\hat{b}^\dagger$. Together, $\hat{b}^\dagger$ and $\hat{b}$ are said to be hermitian adjoints.

[11] See Exercise 3.1 for examples.

Furthermore, if $\hat{A}$ is a hermitian operator, then one can show that

$$\langle A \rangle = \int \psi^* \hat{A}\psi d^3r = \int (\hat{A}\psi)^*\psi d^3r$$
$$= \int \psi(\hat{A}\psi)^* d^3r = \left[\int \psi^*\hat{A}\psi d^3r\right]^* = \langle A \rangle^*. \qquad (3.50)$$

Hence, $\langle A \rangle$ is a real number. This is a significant result. In quantum mechanics, a measurable dynamical variable (known as an observable), such as position, linear and angular momentum, kinetic energy, and total energy, is represented by a hermitian operator $\hat{A}$. Equation (3.50) guarantees that the expectation value $\langle A \rangle$ of any observable A is a real number, consistent with the fact that a measurement of an observable quantity always results in a value that is a real number (as opposed to a complex or imaginary number).

Example 3.3 For the momentum operator $\hat{p}_x = -i\hbar\, d/dx$,

$$\langle p_x \rangle = \int_{-\infty}^{\infty} \psi^* \hat{p}_x \psi dx = -i\hbar \int_{-\infty}^{\infty} \psi^* \frac{d\psi}{dx}dx = -i\hbar \int_{-\infty}^{\infty} \left(-\frac{d\psi^*}{dx}\psi\right) dx$$
$$= \int_{-\infty}^{\infty} (\hat{p}_x\psi)^*\psi dx = \left[\int_{-\infty}^{\infty} \psi^*\hat{p}_x\psi dx\right]^* = \langle p_x \rangle^*.$$

3.3   Eigenvalue Problem and Hilbert Space

For any linear operator $\hat{A}$, there exist a set of numbers $\{a_n\}$ and a set of functions $\{u_n\}$ such that

$$\hat{A}u_n = a_n u_n, \qquad (3.51)$$

which is called an eigenvalue equation. The Schrödinger equation

$$\hat{H}u_n = E_n u_n \qquad (3.52)$$

[12] In fact, the title of Schrödinger's original paper in 1926 was "Quantization as an eigenvalue problem." [E. Schrödinger, *Annalen der Physik* **384**, 361 (1926).]

is a typical example of an eigenvalue equation.[12]

Let us consider two eigenfunctions, u_n and u_m, of a hermitian operator $\hat{A}$ with respective eigenvalues, a_n and a_m. Then, since $\hat{A}u_n = a_n u_n$ and $\hat{A}u_m = a_m u_m$, one can immediately show that

$$\int u_m^* \hat{A} u_n d^3\mathbf{r} = a_n \int u_m^* u_n d^3\mathbf{r}, \tag{3.53}$$

$$\int u_n^* \hat{A} u_m d^3\mathbf{r} = a_m \int u_n^* u_m d^3\mathbf{r}. \tag{3.54}$$

On the other hand, since $\hat{A}$ is hermitian,

$$\int u_n^* \hat{A} u_m d^3\mathbf{r} = \int (\hat{A}u_n)^* u_m d^3\mathbf{r} = a_n^* \int u_n^* u_m d^3\mathbf{r}. \tag{3.55}$$

Therefore, by subtracting Equation (3.55) from Equation (3.54), one can get

$$0 = (a_m - a_n^*) \int u_n^* u_m d^3\mathbf{r}. \tag{3.56}$$

When $n = m$,

$$0 = (a_n - a_n^*) \int |u_n|^2 d^3\mathbf{r} \implies a_n = a_n^*. \tag{3.57}$$

Namely, the eigenvalues of a hermitian operator are real. On the other hand, when $n \neq m$,

$$0 = \int u_m^* u_n d^3\mathbf{r}. \tag{3.58}$$

Assuming that all u_n are normalized, we can conclude that the eigenfunctions u_n of a hermitian operator $\hat{A}$ are orthonormal, i.e.,

$$\boxed{\int u_n^*(\mathbf{r}) u_m(\mathbf{r}) d^3\mathbf{r} = \delta_{nm} = \begin{cases} 1, & n = m, \\ 0, & n \neq m. \end{cases}} \tag{3.59}$$

Example 3.4 For the eigenfunctions, Equation (2.43), for an infinite-barrier 1D quantum well with length L, depicted in Figure 2.5, we have

$$\int_0^L u_n^* u_n dx = \frac{2}{L} \int_0^L \sin^2\left(\frac{n\pi x}{L}\right) dx = 1, \tag{3.60}$$

$$\int_0^L u_n^* u_{m \neq n} dx = \frac{2}{L} \int_0^L \sin\left(\frac{n\pi x}{L}\right) \sin\left(\frac{m\pi x}{L}\right) dx = 0. \tag{3.61}$$

Furthermore, the eigenfunctions $\{u_n(\mathbf{r})\}$ form a complete set in the sense that one can expand an arbitrary function $\psi(\mathbf{r})$ as

$$\psi(\mathbf{r}) = \sum_n b_n u_n(\mathbf{r}). \tag{3.62}$$

Here, b_n are expansion coefficients and can be considered to be the projection of $\psi(r)$ onto $u_n(r)$ because

$$\int u_m^*(r)\psi(r)d^3r = \sum_n b_n \int u_m^*(r)u_n(r)d^3r = b_m. \qquad (3.63)$$

Thus, $u_n(r)$ can be considered to be a "unit vector" in an infinite-dimensional space, called Hilbert space. See Table 3.1 for the correspondence between vector space and Hilbert space.

A vector in an N-dimensional vector space[13] is written as

$$A = A_1 e_1 + A_2 e_2 + \cdots + A_N e_N = \sum_{i=1}^{N} A_i e_i, \qquad (3.64)$$

where the e_i are orthonormal,

$$e_k \cdot e_l = \delta_{kl}, \qquad (3.65)$$

which is analogous to Equation (3.59). The expansion coefficient A_j is the projection of A onto e_j since

$$A_j = e_j \cdot A, \qquad (3.66)$$

which is analogous to Equation (3.63).

In vector space, one can add two vectors to create another vector:

$$A + B = C. \qquad (3.67)$$

Using the expansions

$$A = \sum_i A_i e_i, \qquad (3.68)$$

$$B = \sum_i B_i e_i, \qquad (3.69)$$

one can write

$$C = \sum_i (A_i + B_i)e_i = \sum_i C_i e_i. \qquad (3.70)$$

Similarly, one can make an acceptable wavefunction in Hilbert space by adding two wavefunctions,

$$\phi(r) + \chi(r) = \psi(r); \qquad (3.71)$$

or, using expansions,

$$\phi(r) = \sum_n b_n u_n(r), \qquad (3.72)$$

$$\chi(r) = \sum_n c_n u_n(r), \qquad (3.73)$$

[13] See, e.g., Gilbert Strang, *Introduction to Linear Algebra, Fifth Edition* (Wellesley Cambridge Press, 2016).

Table 3.1 Correspondence between vector space and Hilbert space.

Vector space	Hilbert space
A	$\psi(r)$
e_i	$u_n(r)$
$A = \sum_i A_i e_i$	$\psi(r) = \sum_n b_n u_n(r)$
$e_k \cdot e_l = \delta_{kl}$	$\int u_n^*(r) u_m(r) d^3 r = \delta_{nm}$
$A_j = e_j \cdot A$	$b_n = \int u_n^*(r)\psi(r) d^3 r$
$A + B = C$	$\phi(r) + \chi(r) = \psi(r)$
$A \cdot B = \sum_i A_i B_i$	$\int \phi^*(r)\chi(r) d^3 r = \sum_n b_n^* c_n$
$|A|^2 = \sum_i A_i^2$	$\int |\phi(r)|^2 d^3 r = \sum_n |b_n|^2$

one can write

$$\psi(r) = \sum_n (b_n + c_n)u_n(r) = \sum_n d_n u_n(r). \tag{3.74}$$

In vector space, the inner (or scalar) product of two vectors A and B is given by

$$A \cdot B = \sum_i A_i e_i \cdot \sum_j B_j e_j = \sum_{i,j} A_i B_j \, e_i \cdot e_j \tag{3.75}$$

$$= \sum_{i,j} A_i B_j \delta_{ij} = \sum_i A_i B_i. \tag{3.76}$$

When $A = B$,

$$|A|^2 = A \cdot A = \sum_i A_i^2. \tag{3.77}$$

Analogously, in Hilbert space, the inner (or scalar) product of two wavefunctions $\phi(r)$ and $\chi(r)$ is given by

$$\int \phi^*(r)\chi(r) d^3 r = \int \left(\sum_n b_n u_n(r) \right)^* \left(\sum_m c_m u_m(r) \right) d^3 r \tag{3.78}$$

$$= \sum_{n,m} b_n^* c_m \int u_n^*(r) u_m(r) d^3 r \tag{3.79}$$

$$= \sum_{n,m} b_n^* c_m \delta_{nm} = \sum_n b_n^* c_n. \tag{3.80}$$

When $\phi(r) = \chi(r)$,

$$\int |\phi(r)|^2 d^3 r = \sum_n |b_n|^2. \tag{3.81}$$

Hence, for a normalized wavefunction $\phi(r) = \sum_n b_n u_n(r)$,

$$\sum_n |b_n|^2 = 1 \tag{3.82}$$

must hold.

3.4 Dirac Notation

In this section, we describe a mathematical notation for describing quantum mechanics in a most economical and concise manner, introduced by Paul Dirac.[14] We can rewrite essentially all equations in the previous sections in this chapter using this convenient notation in simpler ways.[15]

In Dirac notation, with each wavefunction $\psi(r)$ we associate a state vector $|\psi\rangle$, called a ket, which lives in a vector space called a ket space. We can add two kets, $|\psi\rangle$ and $|\chi\rangle$, to create another ket, $|\phi\rangle$, in the ket space:

$$|\phi\rangle = |\chi\rangle + |\psi\rangle. \tag{3.83}$$

One can also create a new ket by multiplying a ket by a number, c,

$$|\phi\rangle = c|\psi\rangle, \tag{3.84}$$

or by operating a linear operator $\hat{A}$ (from the left) on a ket,

$$|\phi\rangle = \hat{A}|\psi\rangle. \tag{3.85}$$

Also, we associate the complex conjugate of a wavefunction $\psi^*(r)$ with a quantity $\langle\psi|$, called a bra, which lives in a bra space. We can add two bras, $\langle\psi|$ and $\langle\chi|$, to create another bra, $\langle\phi|$, in the bra space:

$$\langle\phi| = \langle\chi| + \langle\psi|. \tag{3.86}$$

One can also create a new bra by multiplying a bra by a number, c,

$$\langle\phi| = c\langle\psi|, \tag{3.87}$$

or by operating a linear operator $\hat{A}$ on a bra[16]

$$\langle\phi| = \langle\psi|\hat{A}. \tag{3.88}$$

The scalar product of $\psi(r)$ and $\phi^*(r)$ is associated with a bracket:

$$\langle\phi|\psi\rangle := \int \phi^*(r)\psi(r)d^3r. \tag{3.89}$$

It immediately follows that

$$\langle\phi|\psi\rangle^* = \langle\psi|\phi\rangle, \tag{3.90}$$

that is, changing the order of ψ and ϕ in their inner product requires taking the complex conjugate of the bracket. We can also see that

$$\int \phi^*(r)\hat{A}\psi(r)d^3r = \langle\phi|\hat{A}\psi\rangle. \tag{3.91}$$

[14] See P. A. M. Dirac, *The Principles of Quantum Mechanics* (Oxford University Press, 1930).

[15] In consistency with the contemporary literature of quantum science and technology, we will use Dirac notation wherever appropriate throughout the remaining chapters of this book.

[16] Note that an operator acts on a bra *from the right*.

Dual correspondence:
For each ket space, there exists a bra space that is dual to the ket space. Each ket has a unique bra according to the following dual correspondence:

$$|\psi\rangle \Leftrightarrow \langle\psi|, \qquad (3.95)$$

$$c|\psi\rangle \Leftrightarrow c^*\langle\psi|, \qquad (3.96)$$

$$\hat{A}|\psi\rangle \Leftrightarrow \langle\psi|\hat{A}^\dagger, \qquad (3.97)$$

where $\hat{A}^\dagger$ is the hermitian adjoint of $\hat{A}$.

[17] This can be seen as

$$(|\phi\rangle\langle\psi|)\,|\chi\rangle \;=\; \langle\psi|\chi\rangle\,|\phi\rangle,$$
$$\text{(operator) (ket) = (number) (ket)}$$

i.e., $|\phi\rangle\langle\psi|$ changes a ket, $|\chi\rangle$, into another ket, $\langle\psi|\chi\rangle|\phi\rangle$. The operator $|\phi\rangle\langle\psi|$ is also known as the outer product of $|\phi\rangle$ and $|\psi\rangle$.

[18] This can be seen as

$$(\langle\phi|)\,(\hat{A}|\psi\rangle) = (\langle\phi|\hat{A})\,|\psi\rangle,$$
$$\text{(bra) (ket) = (bra) (ket)}$$

From either perspective, the end result is a bracket.

If $\hat{A}$ is a hermitian operator, we can move its position from inside the ket to inside the bra in the following expression:

$$\langle\psi|\hat{A}\phi\rangle = \langle\hat{A}\psi|\phi\rangle. \qquad (3.92)$$

In the case where $\hat{A} = c$ is a number, we can simply take it out of the ket to the front of the bracket as

$$\langle\psi|c\phi\rangle = c\langle\psi|\phi\rangle, \qquad (3.93)$$

but we have to take its complex conjugate when we take it out of the bra:

$$\langle c\psi|\phi\rangle = c^*\langle\psi|\phi\rangle. \qquad (3.94)$$

We have so far seen three types of products: $\hat{A}|\psi\rangle$, which is a ket, $\langle\psi|\hat{A}$, which is a bra, and $\langle\phi|\psi\rangle$, which is a number. Another type of product is $|\phi\rangle\langle\psi|$, which is actually an operator.[17] Further, another interesting type of product of different quantities is $\langle\phi|\hat{A}|\psi\rangle$, which is a bracket (or a bra-ket), i.e., a number.[18] Note that $\langle\phi|\hat{A}|\psi\rangle = \langle\phi|\,(\hat{A}|\psi\rangle) = \{((\langle\psi|\hat{A}^\dagger)\,|\phi\rangle)\}^* = \langle\psi|\hat{A}^\dagger|\phi\rangle^*$. Therefore, if $\hat{A}$ is hermitian (i.e., $\hat{A} = \hat{A}^\dagger$) then $\langle\phi|\hat{A}|\psi\rangle = \langle\psi|\hat{A}|\phi\rangle^* = \langle\psi|\hat{A}\phi\rangle^* = \langle\hat{A}\phi|\psi\rangle$.

Using Dirac notation, an eigenvalue equation is written as

$$\hat{A}|n\rangle = a_n|n\rangle. \qquad (3.98)$$

The orthonormality relationship for the eigenstates $|n\rangle$ of a hermitian operator is written in compact form as

$$\langle n|m\rangle = \delta_{nm}. \qquad (3.99)$$

Any ket $|\psi\rangle$ can be expanded as

$$|\psi\rangle = \sum_n b_n|n\rangle. \qquad (3.100)$$

By multiplying an eigenbra $\langle m|$ from the left, we obtain

$$\langle m|\psi\rangle = \sum_n b_n\langle m|n\rangle = b_m. \qquad (3.101)$$

By substituting Equation (3.101) into Equation (3.100), we get

$$|\psi\rangle = \sum_n \langle n|\psi\rangle|n\rangle = \sum_n |n\rangle\langle n|\psi\rangle. \qquad (3.102)$$

Since $|\psi\rangle$ is an arbitrary ket, we must have

$$\sum_n |n\rangle\langle n| := \hat{1}; \qquad (3.103)$$

the identity operator $\hat{1}$ can be inserted in any place, e.g.,

$$\langle\psi|\psi\rangle = \langle\psi|\left(\sum_n |n\rangle\langle n|\right)|\psi\rangle = \sum_n\langle\psi|n\rangle\langle n|\psi\rangle$$
$$= \sum_n |\langle n|\psi\rangle|^2 = \sum_n |b_n|^2 = 1. \tag{3.104}$$

Also,

$$\langle A\rangle = \langle\psi|\hat{A}|\psi\rangle = \langle\psi|\left(\sum_n |n\rangle\langle n|\right)\hat{A}\left(\sum_m |m\rangle\langle m|\right)|\psi\rangle$$
$$= \sum_{n,m}\langle\psi|n\rangle\langle m|\hat{A}|n\rangle\langle m|\psi\rangle = \sum_{n,m}\langle\psi|n\rangle\, a_n\delta_{mn}\,\langle m|\psi\rangle$$
$$= \sum_n a_n|\langle n|\psi\rangle|^2 = \sum_n a_n|b_n|^2. \tag{3.105}$$

Remark 3.2 The number-state representation

In Dirac notation, the Schrödinger equation for a 1D SHO, Equation (3.26), can be rewritten as $\hat{n}|n\rangle = \varepsilon'_n|n\rangle$. Here $\hat{n} := \hat{b}^\dagger\hat{b}$ is called the number operator, and the kets $|n\rangle$ are called number states (or Fock states in quantum optics[19]). They form a complete orthonormal set:

$$\langle n|n\rangle = \delta_{nm} \quad \text{and} \quad \sum_{n=0}^{\infty}|n\rangle\langle n| = \hat{1} \tag{3.106}$$

[19] See, e.g., A. M. Fox, *Quantum Optics: An Introduction* (Oxford University Press, 2006).

Recalling that $\varepsilon' = \varepsilon - 1/2$ and $\varepsilon = E/\hbar\omega_0$, we can rewrite the SHO Schrödinger equation as

$$\hat{H}_{\text{SHO}}|n\rangle = \hbar\omega_0\left(\hat{n} + \frac{1}{2}\right)|n\rangle = E_n|n\rangle. \tag{3.107}$$

For the ground state $|0\rangle$, $\hat{n}|0\rangle = 0$ since $\hat{b}|0\rangle = 0$ and therefore, $\frac{1}{2}\hbar\omega_0|0\rangle = E_0|0\rangle$, i.e., $E_0 = \frac{1}{2}\hbar\omega_0$. However, by operating $\hat{b}^\dagger$ on Equation (3.107) from the left and using the commutation relation $[\hat{b}, \hat{b}^\dagger] = 1$, we can show that

$$\hbar\omega_0\left(\hat{n} + \frac{1}{2}\right)(\hat{b}^\dagger|n\rangle) = (E_n + \hbar\omega_0)(\hat{b}^\dagger|n\rangle), \tag{3.108}$$

which suggests that $\hat{b}^\dagger|n\rangle$ is an eigenket and $E_1 = E_0 + \hbar\omega_0 = \frac{3}{2}\hbar\omega_0$, $E_2 = E_1 + \hbar\omega_0 = \frac{5}{2}\hbar\omega_0$, ... and

$$E_n = E_0 + n\hbar\omega_0 = \left(n + \frac{1}{2}\right)\hbar\omega_0. \tag{3.109}$$

By combining Equations (3.107) and (3.109), we conclude that

$$\hat{n}|n\rangle = n|n\rangle. \tag{3.110}$$

In other words, the number states are the eigenstates of both $\hat{H}_{\text{SHO}}$ and $\hat{n}$. The number states $|n\rangle$ and $|n+1\rangle$ are related by $\hat{b}^\dagger|n\rangle = \alpha_n|n+1\rangle$, and the coefficient α_n can be determined through

$$|\alpha_n|^2 = \langle n+1|\alpha_n^* \cdot \alpha_n|n+1\rangle = \langle n|\hat{b}\,\hat{b}^\dagger|n\rangle$$
$$= \langle n|(\hat{b}^\dagger\hat{b} + [\hat{b}, \hat{b}^\dagger])|n\rangle = \langle n|(\hat{n}+1)|n\rangle = n+1. \tag{3.111}$$

Hence,

$$\hat{b}^\dagger |n\rangle = \sqrt{n+1}\,|n+1\rangle. \tag{3.112}$$

Similarly, we obtain

$$\hat{b}|n\rangle = \sqrt{n}\,|n-1\rangle. \tag{3.113}$$

Using Equation (3.112) recursively, we obtain

$$
\begin{aligned}
|n\rangle &= \frac{1}{\sqrt{n}}\,\hat{b}^\dagger|n-1\rangle = \frac{1}{\sqrt{n}}\,\hat{b}^\dagger\frac{1}{\sqrt{n-1}}\,\hat{b}^\dagger|n-2\rangle \\
&= \frac{1}{\sqrt{n}}\,\hat{b}^\dagger\frac{1}{\sqrt{n-1}}\,\hat{b}^\dagger\frac{1}{\sqrt{n-2}}\,\hat{b}^\dagger|n-3\rangle = \ldots = \frac{(\hat{b}^\dagger)^n}{\sqrt{n!}}|0\rangle.
\end{aligned} \tag{3.114}
$$

[20] In the quantum theory of light, the coherent state was introduced by Glauber in 1963 and is called the Glauber state. The state was first derived by Schrödinger in 1926 as a minimum-uncertainty Gaussian wavepacket. See Exercise 3.6 for more about properties of the coherent state.

We can define the so-called *coherent state* using number states:[20]

$$|\alpha\rangle := e^{-|\alpha|^2/2}\sum_{n=0}^{\infty}\frac{\alpha^n}{\sqrt{n!}}|n\rangle. \tag{3.115}$$

Here α is a dimensionless complex number. We can show that the coherent state is an eigenstate of the annihilation operator, $\hat{b}$:

$$
\begin{aligned}
\hat{b}|\alpha\rangle &= e^{-|\alpha|^2/2}\sum_{n=0}^{\infty}\frac{\alpha^n}{\sqrt{n!}}\hat{b}|n\rangle = e^{-|\alpha|^2/2}\sum_{n=0}^{\infty}\frac{\alpha^n}{\sqrt{n!}}\sqrt{n}\,|n-1\rangle \\
&= \alpha\,e^{-|\alpha|^2/2}\sum_{n=0}^{\infty}\frac{\alpha^n}{\sqrt{n!}}\hat{b}|n\rangle = \alpha|\alpha\rangle
\end{aligned} \tag{3.116}
$$

The coherent state is a minimum-uncertainty state and describes light generated by a highly stabilized laser well above the threshold.

3.5 Superposition and Quantum Measurement

Quantum superposition and quantum measurement are two of the concepts that make quantum mechanics drastically different from classical mechanics – they are so nonintuitive and foreign to our conception of the classical, macroscopic world. Here, we discuss these nonclassical concepts through the expansion theorem and Heisenberg's uncertainty principle.

3.5.1 Interpretation of the Expansion Coefficients

In this section, we discuss the physical meaning of the expansion coefficients, b_n, appearing in Equations (3.104) and (3.105). This is deeply related to the intrinsically probabilistic nature of quantum mechanics as well as to the subtle role of quantum measurement, a subject that caused many important debates among some of the most influential founders of quantum mechanics, such as Bohr and

Einstein. Furthermore, quantum superposition and quantum measurement play central roles in quantum information processing, as described in detail in Chapter 4.

Suppose that a certain observable A is represented by a hermitian operator $\hat{A}$, which has a set of eigenkets $\{|n\rangle\}$ and corresponding eigenvalues $\{a_n\}$: $\hat{A}|n\rangle = a_n|n\rangle$. Suppose now that the observable A is measured for a collection of systems, each of which is described by the same wavefunction $|\psi\rangle = \sum_n b_n|n\rangle$. There are three important hypotheses regarding this situation (see the box on the left), which have been experimentally verified.

In order to understand these hypotheses more clearly, let us consider, as the simplest example, a two-level system consisting of eigenstates $|1\rangle$ and $|2\rangle$ with eigenvalues a_1 and a_2, respectively. The states $|1\rangle$ and $|2\rangle$ form a complete orthonormal set. Thus, using Equation (3.103), we can write

$$|\psi\rangle = \hat{1}|\psi\rangle = \left(\sum_{n=1}^{2} |n\rangle\langle n|\right)|\psi\rangle$$
$$= |1\rangle\langle 1|\psi\rangle + |2\rangle\langle 2|\psi\rangle = b_1|1\rangle + b_2|2\rangle. \qquad (3.117)$$

The bra corresponding to $|\psi\rangle$ is

$$\langle\psi| = b_1^*\langle 1| + b_2^*\langle 2|. \qquad (3.118)$$

Therefore,

$$\langle\psi|\psi\rangle = \left(b_1^*\langle 1| + b_2^*\langle 2|\right)\left(b_1|1\rangle + b_2|2\rangle\right)$$
$$= |b_1|^2 + |b_2|^2 = 1. \qquad (3.119)$$

The expectation value of A is calculated as follows:[21]

$$\langle A\rangle = \langle\psi|\hat{A}|\psi\rangle$$
$$= \left(b_1^*\langle 1| + b_2^*\langle 2|\right)\hat{A}\left(b_1|1\rangle + b_2|2\rangle\right)$$
$$= \left(b_1^*\langle 1| + b_2^*\langle 2|\right)\left(a_1 b_1|1\rangle + a_2 b_2|2\rangle\right)$$
$$= a_1|b_1|^2 + a_2|b_2|^2. \qquad (3.120)$$

If $|\psi\rangle$ happens to be $|1\rangle$, i.e., $b_1 = 1$ and $b_2 = 0$, then we can see that $\langle A\rangle = a_1$ with certainty (100% probability). If $b_1 = b_2 = \frac{1}{\sqrt{2}}$ then, on average, the measurement of A results in a_1 for half the time and a_2 for the other half.[22]

Three hypotheses regarding superposition and measurement:

1. The result of measurement can only be one of the eigenvalues a_n.

2. The probability that the eigenvalue a_n will be found is $|b_n|^2$.

3. After a measurement yields an eigenvalue a_1, for example, then the system must be in the state $|1\rangle$.

[21] This can be understood to be the "weighted average" of the two possible values, a_1 and a_2.

[22] Quantum measurement does not allow us to predict with certainty which value will result for a given particle.

The third hypothesis regarding superposition and measurement hints at the active role measurement plays in quantum mechanics.[23] After a measurement yields an eigenvalue a_n, the system must be in the state $|n\rangle$. Suppose that the two-level system under consideration before measurement is in the superposition state

$$\frac{1}{\sqrt{2}}(|1\rangle + |2\rangle). \tag{3.121}$$

Namely, *prior to measurement*, the quantum state of the system is described as a 50%–50% linear combination of $|1\rangle$ and $|2\rangle$. On the other hand, *after measurement*, the system is in one, and only one, of the eigenstates ($|1\rangle$ or $|2\rangle$). Therefore, the act of quantum measurement changed two *possibilities* ($|1\rangle$ and $|2\rangle$) to one *actuality* ($|1\rangle$ or $|2\rangle$). The value of A is completely *undetermined* prior to measurement. Quantum measurement *projects* a previously undetermined quantum state into one of the eigenstates. This projection process, occurring as a result of quantum measurement, is called wavefunction collapse.

3.5.2   *The Uncertainty Principle*

The above discussion highlights the fragile nature of quantum systems. That is, measurement can significantly influence, or disturb, a quantum system. In order to evaluate this aspect more quantitatively, let us consider a thought experiment, à la Heisenberg. See Figure 3.2. We want to determine the position x of a particle precisely using an optical microscope. We illuminate the particle with a light beam and measure the scattered light with a camera. The position uncertainty will be given by the spatial resolution of the system, which is limited to be $\Delta x \sim \lambda/2\text{NA}$,[24] where λ is the wavelength of the light used, and $\text{NA} = \sin\theta$ is the numerical aperture of the microscope; θ is the light scattering angle. If we want to further increase the precision of the position measurement (i.e., decrease Δx), we have to shorten the wavelength, but that means that the photon energy $h\nu = hc/\lambda$ increases. Now remember the Compton effect:[25] the particle will be in collision with a high-energy photon and acquire a large momentum, whose x component p_x can vary from $-(h/\lambda)\sin\theta$ to $+(h/\lambda)\sin\theta$. In other words, $\Delta p_x \sim (2h/\lambda)\sin\theta$, and therefore, $\Delta x \, \Delta p_x \sim h$. This is the essence of the uncertainty principle.

So x and p_x cannot be determined with certainty simultaneously, but would this conclusion apply to any two observables? If not, what are the conditions for two observables to be simultaneously determinable precisely? To answer these questions, let us consider two observables, A and B, represented by hermitian operators, $\hat{A}$ and $\hat{B}$,

[23] This was another subject of intense debate in the early days of quantum mechanics. Einstein and Schrödinger thought that each particle is already in a predetermined state ($|1\rangle$ or $|2\rangle$) but that we simply do not know which state. Namely, they argued that measurement, which is classical, merely tells us which state. This interpretation directly went against the Copenhagen interpretation of Bohr, Heisenberg, Pauli, and others, which has proven to be correct, although the concept of wavefunction collapse has made many physicists uneasy. See, e.g., a recent paper by K.-W. Bong *et al.*, *Nature Physics* **16**, 1199 (2020), and references cited therein.

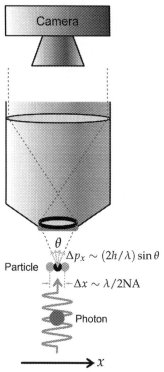

Figure 3.2 Heisenberg's thought experiment illuminating the concept of the uncertainty principle. © Deyin Kong (Rice University).

[24] This formula is known as Abbe's diffraction limit.

[25] See Exercise 2.3.

respectively, with corresponding eigenvalues $\{a_n\}$ and $\{b_n\}$. A measurement of A must result in a number that is a member of $\{a_n\}$, while a measurement of B must result in a number that is a member of $\{b_n\}$. Since each measurement disturbs the system, we can expect that the *order* of the two measurements is important in determining what results are obtained. More specifically, we can imagine that the order would matter when the two operators do not commute ($[\hat{A}, \hat{B}] \neq 0$), such as $\hat{x}$ and $\hat{p}_x$.[26] Measurements of A and B will then interfere with and influence each other so these observables cannot be determined simultaneously[27] with arbitrary accuracy.

Let us assume the special case in which $\{|n\rangle\}$ are simultaneous eigenstates of $\hat{A}$ and $\hat{B}$, i.e., $\hat{A}|n\rangle = a_n|n\rangle$ and $\hat{B}|n\rangle = b_n|n\rangle$. By operating $\hat{B}$ and $\hat{A}$ on these equations, respectively, we get $\hat{B}\hat{A}|n\rangle = a_n\hat{B}|n\rangle = a_nb_n|n\rangle$ and $\hat{A}\hat{B}|n\rangle = b_n\hat{A}|n\rangle = a_nb_n|n\rangle$, which means that $\hat{A}\hat{B}|n\rangle = \hat{B}\hat{A}|n\rangle$, or

$$(\hat{A}\hat{B} - \hat{B}\hat{A})|n\rangle = [\hat{A}, \hat{B}]|n\rangle = 0. \tag{3.122}$$

Since this holds for the complete set $\{|n\rangle\}$,

$$[\hat{A}, \hat{B}]|\psi\rangle = [\hat{A}, \hat{B}]\sum_n c_n|n\rangle = \sum_n c_n[\hat{A}, \hat{B}]|n\rangle = 0 \tag{3.123}$$

for any $|\psi\rangle = \sum_n c_n|n\rangle$. Therefore, we conclude that $[\hat{A}, \hat{B}] = 0$.

Conversely, if two hermitian operators $\hat{A}$ and $\hat{B}$ commute ($\hat{A}\hat{B} = \hat{B}\hat{A}$), then for the eigenstates $\{|n^{(A)}\rangle\}$ of $\hat{A}$

$$\hat{A}\hat{B}|n^{(A)}\rangle = \hat{B}\hat{A}|n^{(A)}\rangle = a_n\hat{B}|n^{(A)}\rangle \tag{3.124}$$

or

$$\hat{A}(\hat{B}|n^{(A)}\rangle) = a_n(\hat{B}|n^{(A)}\rangle). \tag{3.125}$$

Hence, $\hat{B}|n^{(A)}\rangle$ is an eigenfunction of $\hat{A}$ with eigenvalue a_n. If there is only one eigenstate of $\hat{A}$ corresponding to the eigenvalue a_n, then this implies that $\hat{B}|n^{(A)}\rangle$ must be proportional to $|n^{(A)}\rangle$, that is,

$$\hat{B}|n^{(A)}\rangle = b_n|n^{(A)}\rangle. \tag{3.126}$$

Then, $|n^{(A)}\rangle$ (for any n) is a simultaneous eigenstate of $\hat{A}$ and $\hat{B}$.

What would happen if we perform sequential measurements of observables A and B when $\hat{A}$ and $\hat{B}$ commute? In this case, the two measurements will not interfere with each other, and thus, it should be possible to determine both quantities with arbitrary certainty. Suppose one measures B first and gets b_m, which is one of the eigenvalues of $\hat{B}$. After measurement, the system is "left" in the eigenstate $|m\rangle$. However, since $\hat{A}$ and $\hat{B}$ commute, this is also an

[26] See Remark 3.1.

[27] That is, one immediately after the other.

eigenstate of $\hat{A}$, i.e., $\hat{A}|m\rangle = a_m|m\rangle$. Thus, a subsequent measurement of A will yield, *with certainty*, the value a_m. One can go back and measure B again and obtain the same value b_m, *with certainty*. We can thus make a simultaneous determination of A and B.[28]

Now consider performing sequential measurements of A and B when $\hat{A}$ and $\hat{B}$ do *not* commute. A measurement of A yields a_m (an eigenvalue of $\hat{A}$), leaving the system in $|m^{(A)}\rangle$ (an eigenstate of $\hat{A}$). In this case, however, $|m^{(A)}\rangle$ is *not* an eigenstate of $\hat{B}$, in general. Thus, we have to expand $|m^{(A)}\rangle$ in terms of the eigenstates of $\hat{B}$:

$$|m^{(A)}\rangle = \sum_j c_j|j^{(B)}\rangle, \tag{3.127}$$

$$\sum_j |c_j|^2 = 1. \tag{3.128}$$

Now if one measures B, it can result in any member of $\{b_n\}$, but with a corresponding probability $|c_n|^2 < 1$. Namely, measuring A introduced an uncertainty into the measurement of B. Therefore, A and B cannot be measured simultaneously with certainty.

To conclude this section, we provide a mathematical proof of the uncertainty principle. For two observables A and B and a state $|\psi\rangle$, we define the following two operators:

$$\Delta\hat{A} := \hat{A} - \langle A\rangle, \tag{3.129}$$

$$\Delta\hat{B} := \hat{B} - \langle B\rangle, \tag{3.130}$$

where $\langle A\rangle = \langle\psi|A|\psi\rangle$ and $\langle B\rangle = \langle\psi|B|\psi\rangle$. We also define

$$|\chi\rangle := \Delta\hat{A}|\psi\rangle, \tag{3.131}$$

$$|\phi\rangle := \Delta\hat{B}|\psi\rangle. \tag{3.132}$$

By further defining the uncertainty to be the standard deviation from the expectation value,[29]

$$\Delta A := \sqrt{\langle(A - \langle A\rangle)^2\rangle} = \sqrt{\langle A^2\rangle - \langle A\rangle^2}, \tag{3.133}$$

$$\Delta B := \sqrt{\langle(B - \langle B\rangle)^2\rangle} = \sqrt{\langle B^2\rangle - \langle B\rangle^2}, \tag{3.134}$$

we can show that

$$\langle\chi|\chi\rangle = (\Delta A)^2, \tag{3.135}$$

$$\langle\phi|\phi\rangle = (\Delta B)^2. \tag{3.136}$$

In addition, we can obtain the following relationship:

$$\langle\chi|\phi\rangle = \langle\psi|\Delta\hat{A}\Delta\hat{B}|\psi\rangle = \langle\Delta\hat{A}\Delta\hat{B}\rangle$$
$$= \frac{1}{2}\langle[\hat{A},\hat{B}]\rangle + \frac{1}{2}\langle\{\hat{A},\hat{B}\}\rangle, \tag{3.137}$$

where

$$\{\hat{A}, \hat{B}\} := \hat{A}\hat{B} + \hat{B}\hat{A} \tag{3.138}$$

is known as the anticommutator of $\hat{A}$ and $\hat{B}$. By substituting $\langle \chi | \chi \rangle$, $\langle \phi | \phi \rangle$, and $\langle \chi | \phi \rangle$ into the Cauchy–Schwarz inequality[30]

$$\langle \chi | \chi \rangle \langle \phi | \phi \rangle \geq |\langle \chi | \phi \rangle|^2, \tag{3.139}$$

we obtain

$$(\Delta A)^2 (\Delta B)^2 \geq \frac{1}{4} |\langle [\hat{A}, \hat{B}] \rangle|^2 + \frac{1}{4} |\langle \{\hat{A}, \hat{B}\} \rangle|^2. \tag{3.140}$$

Here we made use of the fact that $[\hat{A}, \hat{B}]$ is antihermitian[31] (and thus its expectation value is imaginary) and $\{\hat{A}, \hat{B}\}$ is hermitian (and thus its expectation value is real). Since the second term on the right-hand side of Equation (3.140) is a positive real number, we get

$$\boxed{\Delta A \, \Delta B \geq \frac{1}{2} |\langle [\hat{A}, \hat{B}] \rangle|.} \tag{3.141}$$

Because $[\hat{x}, \hat{p}_x] = i\hbar$, we obtain

$$\Delta x \, \Delta p_x \geq \frac{\hbar}{2}. \tag{3.142}$$

3.6 Matrix Formulation

An alternative to Schrödinger's formulation of quantum mechanics (*wave mechanics*) is the matrix formulation of quantum mechanics, or *matrix mechanics*, developed by Heisenberg, Born, and Jordan.[32] In this section, we combine the mathematical methods and tools for treating quantum states and operators in Hilbert space using Dirac notation, as detailed in the previous sections in this chapter, in order to introduce the basics of matrix mechanics.

When a complete set of states, $\{|n\rangle\}$, is given as the basis vectors for a Hilbert space, an operator $\hat{A}$ in that space is represented by a matrix defined by the following (i, j) element:

$$(\mathbf{A})_{ij} := \langle i | \hat{A} | j \rangle. \tag{3.143}$$

With this definition, we can calculate the (i, j) element of the matrix representing the product of two operators, $\hat{A}$ and $\hat{B}$, as

$$(\mathbf{AB})_{ij} = \langle i | \hat{A} \hat{B} | j \rangle = \langle i | \hat{A} \left(\sum_k |k\rangle \langle k| \right) \hat{B} | j \rangle \tag{3.144}$$

$$= \sum_k \langle i | \hat{A} | k \rangle \langle k | \hat{B} | j \rangle = \sum_k (\mathbf{A})_{ik} (\mathbf{B})_{kj}. \tag{3.145}$$

[30] For ordinary vectors $\boldsymbol{u}$ and $\boldsymbol{v}$, the Cauchy–Schwarz inequality is $|\boldsymbol{u}|^2 |\boldsymbol{v}|^2 \geq |\boldsymbol{u} \cdot \boldsymbol{v}|^2$.

[31] See Exercise 3.1.

[32] W. Heisenberg, *Zeitschrift für Physik* **33**, 879 (1925); M. Born and P. Jordan, *ibid.* **34**, 858 (1925); M. Born, W. Heisenberg, and P. Jordan, *ibid* **35**, 557 (1925).

The (i, j) element of the matrix representing the hermitian conjugate of $\hat{A}$ can be obtained as $(\mathbf{A}^{\dagger})_{ij} = \langle i|\hat{A}^{\dagger}|j\rangle = \langle \hat{A}i|j\rangle = \langle j|\hat{A}|i\rangle^* = (\mathbf{A})_{ji}^*$. When the eigenstates of $\hat{A}$ are used as the basis, $(\mathbf{A})_{ij} = \langle i|\hat{A}|j\rangle = a_j\langle i|j\rangle = a_j\delta_{ji}$, that is, $\mathbf{A}$ is diagonal.

Let us now consider a Hilbert space for which the basis vectors are $\{|u_n\rangle\}$. An operator $\hat{A}$ is then represented by a matrix given by $(\mathbf{A})_{ij} = \langle u_i|\hat{A}|u_j\rangle$. Suppose that the eigenstates and eigenvalues of $\hat{A}$ are $\{|a_n\rangle\}$ and $\{a_n\}$, respectively. Each eigenstate of $\hat{A}$ can be expanded in terms of $\{|u_n\rangle\}$:

$$|a_i\rangle = \sum_k |u_k\rangle\langle u_k|a_i\rangle, \tag{3.146}$$

$$\langle a_j| = \sum_l \langle a_j|u_l\rangle\langle u_l|. \tag{3.147}$$

Therefore,

$$\langle a_i|\hat{A}|a_j\rangle = a_i\delta_{ij} = \sum_{k,l}\langle a_i|u_l\rangle\langle u_l|\hat{A}|u_k\rangle\langle u_k|a_j\rangle. \tag{3.148}$$

Now we define a matrix $\mathbf{U}$ as

$$(\mathbf{U})_{kj} := \langle u_k|a_j\rangle. \tag{3.149}$$

Then,

$$\langle a_i|u_l\rangle = \langle u_l|a_i\rangle^* = (\mathbf{U})_{li}^* = (\mathbf{U}^{\dagger})_{il}. \tag{3.150}$$

Thus, we can rewrite Equation (3.148) as

$$a_j\delta_{ij} = \sum_{k,l}(\mathbf{U}^{\dagger})_{ik}(\mathbf{A})_{kl}(\mathbf{U})_{lj} = (\mathbf{U}^{\dagger}\mathbf{A}\mathbf{U})_{ij}. \tag{3.151}$$

Hence, the matrix $(\mathbf{U}^{\dagger}\mathbf{A}\mathbf{U})_{ij}$ is diagonal even though $(\mathbf{A})_{kl}$ is not. The matrix $\mathbf{U}$ is unitary because

$$(\mathbf{U}^{\dagger}\mathbf{U})_{ij} = \sum_k (\mathbf{U}^{\dagger})_{ik}(\mathbf{U})_{kj} = \sum_k \langle a_i|u_k\rangle\langle u_k|a_j\rangle = \delta_{ij}. \tag{3.152}$$

Namely,

$$\mathbf{U}^{\dagger}\mathbf{U} = \mathbf{I} \quad \text{or} \quad \mathbf{U}^{\dagger} = \mathbf{U}^{-1}. \tag{3.153}$$

Finally, we consider a two-level system as an example. Suppose that $\hat{A}$ is a hermitian operator with two eigenstates:

$$\hat{A}|0\rangle = a_0|0\rangle, \quad \hat{A}|1\rangle = a_1|1\rangle. \tag{3.154}$$

We take $\{|0\rangle, |1\rangle\}$ as the basis set to construct a 2D Hilbert space. They can be expressed as column vectors,

$$|0\rangle = \begin{bmatrix} 1 \\ 0 \end{bmatrix}, \quad |1\rangle = \begin{bmatrix} 0 \\ 1 \end{bmatrix}. \tag{3.155}$$

and the corresponding basis bras are expressed as row vectors:

$$\langle 0| = \begin{bmatrix} 1 & 0 \end{bmatrix}, \quad \langle 1| = \begin{bmatrix} 0 & 1 \end{bmatrix}. \tag{3.156}$$

They are orthonormal:

$$\langle 0|0 \rangle = \begin{bmatrix} 1 & 0 \end{bmatrix} \begin{bmatrix} 1 \\ 0 \end{bmatrix} = 1^2 + 0^2 = 1, \tag{3.157}$$

$$\langle 1|1 \rangle = \begin{bmatrix} 0 & 1 \end{bmatrix} \begin{bmatrix} 0 \\ 1 \end{bmatrix} = 0^2 + 1^2 = 1, \tag{3.158}$$

$$\langle 0|1 \rangle = \begin{bmatrix} 1 & 0 \end{bmatrix} \begin{bmatrix} 0 \\ 1 \end{bmatrix} = 1 \times 0 + 0 \times 1 = 0, \tag{3.159}$$

$$\langle 1|0 \rangle = \begin{bmatrix} 0 & 1 \end{bmatrix} \begin{bmatrix} 1 \\ 0 \end{bmatrix} = 0 \times 1 + 1 \times 0 = 0. \tag{3.160}$$

They form a complete set:

$$|0\rangle\langle 0| + |1\rangle\langle 1| = \begin{bmatrix} 1 \\ 0 \end{bmatrix} \begin{bmatrix} 1 & 0 \end{bmatrix} + \begin{bmatrix} 1 \\ 0 \end{bmatrix} \begin{bmatrix} 1 & 0 \end{bmatrix} \tag{3.161}$$

$$= \begin{bmatrix} 1 & 0 \\ 0 & 0 \end{bmatrix} + \begin{bmatrix} 0 & 0 \\ 0 & 1 \end{bmatrix} = \begin{bmatrix} 1 & 0 \\ 0 & 1 \end{bmatrix} = \mathbf{I}_{2D}, \tag{3.162}$$

where $\mathbf{I}_{2D}$ is the 2D unit matrix.

Any ket $|\psi\rangle$ (bra $\langle\psi|$) in this space can be expressed as a 2D column (row) vector, i.e.,

$$|\psi\rangle = \alpha|0\rangle + \beta|1\rangle = \begin{bmatrix} \alpha \\ \beta \end{bmatrix}, \tag{3.163}$$

$$\langle\psi| = \alpha^*\langle 0| + \beta^*\langle 1| = \begin{bmatrix} \alpha^* & \beta^* \end{bmatrix}, \tag{3.164}$$

and its norm is calculated to be

$$\langle\psi|\psi\rangle = \begin{bmatrix} \alpha^* & \beta^* \end{bmatrix} \begin{bmatrix} \alpha \\ \beta \end{bmatrix} = |\alpha|^2 + |\beta|^2. \tag{3.165}$$

A general operator $\hat{B}$ is represented by a matrix as

$$\mathbf{B} = \begin{bmatrix} \langle 0|\hat{B}|0\rangle & \langle 0|\hat{B}|1\rangle \\ \langle 1|\hat{B}|0\rangle & \langle 1|\hat{B}|1\rangle \end{bmatrix}. \tag{3.166}$$

However, because of Equation (3.154), the matrix representing the operator $\hat{A}$ is diagonal in this basis, with the diagonal elements equal to the eigenvalues:

$$\mathbf{A} = \begin{bmatrix} \langle 0|\hat{A}|0\rangle & \langle 0|\hat{A}|1\rangle \\ \langle 1|\hat{A}|0\rangle & \langle 1|\hat{A}|1\rangle \end{bmatrix} = \begin{bmatrix} a_0 & 0 \\ 0 & a_1 \end{bmatrix}. \tag{3.167}$$

In the same basis as above, we can also represent the operators $\hat{A}_+ := |1\rangle\langle 0|$ and $\hat{A}_- := |0\rangle\langle 1|$ as the following matrices:

$$\mathbf{A}_+ = \begin{bmatrix} \langle 0|1\rangle\langle 0|0\rangle & \langle 0|1\rangle\langle 0|1\rangle \\ \langle 1|1\rangle\langle 0|0\rangle & \langle 1|1\rangle\langle 0|1\rangle \end{bmatrix} = \begin{bmatrix} 0 & 0 \\ 1 & 0 \end{bmatrix}, \tag{3.168}$$

$$\mathbf{A}_- = \begin{bmatrix} \langle 0|0\rangle\langle 1|0\rangle & \langle 0|0\rangle\langle 1|1\rangle \\ \langle 1|0\rangle\langle 1|0\rangle & \langle 1|0\rangle\langle 1|1\rangle \end{bmatrix} = \begin{bmatrix} 0 & 1 \\ 0 & 0 \end{bmatrix}. \tag{3.169}$$

3.7 Chapter Summary

In this chapter, we provided a detailed review of the mathematical and conceptual foundations of quantum mechanics. We introduced Dirac notation, which helps avoid mathematical complexities and will be used throughout this book. Linear operators are ubiquitous in quantum mechanics, and it is thus essential to become highly skilled with their basic algebra. As an excellent analytically solvable example, we examined the problem of the 1D simple harmonic oscillator extensively, using operators, and obtained the eigenfunctions and eigenenergies. We introduced the concept of a Hilbert space to view general wavefunctions (states) as "vectors" that can be expanded in terms of basis vectors (or eigenkets). We saw that the probabilistic meaning of the expansion coefficients is deeply related to the meaning of quantum measurement on superposition states. Importantly, the fact that measurement disturbs quantum systems has far-reaching consequences, including the phenomenon of wavefunction collapse, and is behind the uncertainty principle. We showed mathematically how the commutation relationships of observable operators are related to the uncertainty principle. Finally, matrix formulation was introduced, which will be directly useful for quantum information processing, to be covered in the next chapter.

3.8   Exercises

Exercise 3.1 (Hermiticity)

(a) Show that $\hat{A} = -i\partial/\partial x$ is hermitian.

(b) Show that the components of the orbital angular momentum operator $\hat{\boldsymbol{L}} = \hat{\boldsymbol{r}} \times \hat{\boldsymbol{p}}$ are hermitian.

(c) Show that $(\hat{A}^\dagger)^\dagger = \hat{A}$ for any linear operator $\hat{A}$.

(d) Show that $(\hat{A}\hat{B})^\dagger = \hat{B}^\dagger \hat{A}^\dagger$ for any linear operators $\hat{A}$ and $\hat{B}$.

(e) Show that, for any linear operator $\hat{A}$, the following three operators are hermitian: $\hat{B} = \hat{A} + \hat{A}^\dagger$, $\hat{C} = i(\hat{A} - \hat{A}^\dagger)$, and $\hat{D} = \hat{A}\hat{A}^\dagger$.

(f) Show that $\hat{C} = \hat{A}\hat{B}$ does not have to be hermitian even when $\hat{A}$ and $\hat{B}$ are hermitian.

(g) Show that $\hat{n} = \hat{b}^\dagger \hat{b}$ is hermitian.

(h) Show that the commutator of two hermitian operators is antihermitian. Namely, show that $\hat{C}^\dagger = -\hat{C}$ when $\hat{C} = [\hat{A}, \hat{B}]$, where $\hat{A}$ and $\hat{B}$ are hermitian.

(i) Show that an arbitrary operator $\hat{F}$ can be decomposed as $\hat{F} = \hat{F}_+ + i\hat{F}_-$, where $\hat{F}_+ = (\hat{F} + \hat{F}^\dagger)/2$ and $\hat{F}_- = (\hat{F} - \hat{F}^\dagger)/2i$ and both $\hat{F}_+$ and $\hat{F}_-$ are hermitian operators.

Exercise 3.2 (Commutators)

(a) Show that the following relationship holds for any linear operators $\hat{A}$, $\hat{B}$, and $\hat{C}$: $[\hat{A}\hat{B}, \hat{C}] = \hat{A}[\hat{B}, \hat{C}] + [\hat{A}, \hat{C}]\hat{B}$.

(b) Show that $[\hat{A}, \hat{B}]^\dagger = [\hat{B}^\dagger, \hat{A}^\dagger]$.

(c) Show that $[\hat{x}, \hat{p}_x] = i\hbar$.

(d) Show that $[\hat{x}^m, \hat{p}_x] = im\hbar\hat{x}^{m-1}$, where m is a positive integer.

(e) Show that, for any Hamiltonian $\hat{H} = \hat{p}_x^2/2m + V(\boldsymbol{r})$,

$$\hat{p}_x = \frac{im}{\hbar}[\hat{H}, \hat{x}]. \tag{3.170}$$

(f) Calculate $[\hat{b}, \hat{b}^\dagger]$, $[\hat{b}, \hat{b}]$, $[\hat{b}^\dagger, \hat{b}^\dagger]$, $[\hat{n}, \hat{b}]$, and $[\hat{n}, \hat{b}^\dagger]$, where $\hat{b}^\dagger$ and $\hat{b}$ are the SHO creation and annihilation operators, respectively, and $\hat{n} = \hat{b}^\dagger \hat{b}$ is the number operator.

Exercise 3.3 (Bra and ket algebra)

(a) Consider two states $|\psi_1\rangle = 2i|\phi_1\rangle + |\phi_2\rangle - a|\phi_3\rangle + 4|\phi_4\rangle$ and $|\psi_2\rangle = 3|\phi_1\rangle - i|\phi_2\rangle + 5|\phi_3\rangle - |\phi_4\rangle$, where $|\phi_1\rangle$, $|\phi_2\rangle$, $|\phi_3\rangle$, and $|\phi_4\rangle$ are orthonormal kets and a is a constant. Find the value of a such that $|\psi_1\rangle$ and $|\psi_2\rangle$ are orthogonal.

(b) Consider two states $|\psi\rangle = 9i|\phi_1\rangle + 2|\phi_2\rangle$ and $|\chi\rangle = -\frac{i}{\sqrt{2}}|\phi_1\rangle + \frac{1}{\sqrt{2}}|\phi_2\rangle$, where $|\phi_1\rangle$ and $|\phi_2\rangle$ are two orthonormal eigenstates of a certain operator.

   (i) Calculate $\langle\psi|\psi\rangle$, $\langle\chi|\chi\rangle$, and $\langle\psi+\chi|\psi+\chi\rangle$.

   (ii) Calculate $\langle\psi|\chi\rangle$ and $\langle\chi|\psi\rangle$. Are they equal?

   (iii) Calculate $|\psi\rangle\langle\chi|$ and $|\chi\rangle\langle\psi|$. Are they equal?

   (iv) Find the hermitian conjugates of $|\psi\rangle$, $|\chi\rangle$, $|\psi\rangle\langle\chi|$, and $|\chi\rangle\langle\psi|$.

(c) Consider a state $|\psi\rangle = \frac{1}{\sqrt{2}}|\phi_1\rangle + \frac{1}{2}|\phi_2\rangle + \frac{1}{2}|\phi_3\rangle$, where $|\phi_1\rangle$, $|\phi_2\rangle$, and $|\phi_3\rangle$ are three orthonormal eigenstates of an operator $\hat{B}$ such that $\hat{B}|\phi_n\rangle = n^2|\phi_n\rangle$. Find the expectation value of B when the particle under consideration is in the state $|\psi\rangle$.

Exercise 3.4 (Superposition state)

An electron in an infinite 1D square quantum well of width L centered at $x = L/2$ is in a superposition state of the ground and first excited states:[33]

$$|\Psi(x,t)\rangle = \frac{1}{\sqrt{2}}\left[|\Psi_1(x,t)\rangle + |\Psi_2(x,t)\rangle\right]. \qquad (3.171)$$

[33] See Figure 2.4, Equation (2.43), and Remark 2.1.

Find expressions for:

(a) the probability density, $|\Psi(x,t)|^2$;

(b) the average particle position, $\langle x(t)\rangle$; and

(c) the average momentum, $\langle p_x(t)\rangle$.

Exercise 3.5 (Uncertainty relation)

(a) Show that, for any observable A and any normalized state $|\psi\rangle$, the following holds:

$$(\Delta A)^2 := \langle(A - \langle A\rangle)^2\rangle = \langle A^2\rangle - \langle A\rangle^2. \qquad (3.172)$$

Now, the state of a certain quantum system is given by the following real-space wavefunction:

$$|\psi(x)\rangle = \frac{1}{(\sqrt{\pi}a)^{1/2}}\exp\left(-\frac{x^2}{2a^2} + ikx\right), \qquad (3.173)$$

where a is a positive constant.

(b) Compute $\langle x \rangle$, $\langle x^2 \rangle$, and $\langle p \rangle$.

(c) Use the results from (a) and (b) to verify that $\Delta x \, \Delta p = \hbar/2$.

Exercise 3.6 (The coherent state)

For the coherent state (or Glauber state) $|\alpha\rangle$ defined by Equation (3.115), show the following:

(a) $\langle \alpha | \alpha \rangle = 1$.

(b) $\langle \alpha | \hat{b}^\dagger = \alpha^*$.

(c) $|\alpha|^2 = \bar{n}$, where $\bar{n} := \langle n \rangle = \langle \alpha | \hat{n} | \alpha \rangle$.

(d) $(\Delta n)^2 = \bar{n}$ (or $\Delta n = \sqrt{\bar{n}}$), where $(\Delta n)^2 = \langle (n - \bar{n})^2 \rangle$.

(e) $|\langle n | \alpha \rangle|^2 = e^{-|\alpha|^2}(|\alpha|^{2n}/n!) = e^{-\bar{n}}(\bar{n}^n/n!)$.

(f) $\Delta X_1 = \Delta X_2 = \frac{1}{2}$, where $\hat{X}_1 = \frac{1}{2}(\hat{b}^\dagger + \hat{b})$, $\hat{X}_2 = \frac{i}{2}(\hat{b}^\dagger - \hat{b})$, $\Delta X_j = \sqrt{\langle (X_j - \langle X_j \rangle)^2 \rangle} = \sqrt{\langle X_j^2 \rangle - \langle X_j \rangle^2}$, and $\langle X_j \rangle = \langle \alpha | \hat{X}_j | \alpha \rangle$.

Exercise 3.7 (Superposition and measurement)

A hermitian operator $\hat{A}$ representing an observable A is known to have two eigenstates $|\phi_1\rangle$ and $|\phi_2\rangle$, with eigenvalues a_1 and a_2, respectively. Another hermitian operator $\hat{B}$, representing an observable B, is known to have two eigenstates $|\chi_1\rangle$ and $|\chi_2\rangle$, with eigenvalues b_1 and b_2, respectively. Furthermore, it is known that these two sets of eigenstates are related to each other in the following manner:

$$|\phi_1\rangle = \frac{2|\chi_1\rangle + 3|\chi_2\rangle}{\sqrt{13}}, \tag{3.174}$$

$$|\phi_2\rangle = \frac{3|\chi_1\rangle - 2|\chi_2\rangle}{\sqrt{13}}. \tag{3.175}$$

Now suppose that a measurement of A yields a_1. If B is then measured and after that A is measured again, what would be the probability of obtaining a_1 again?

Exercise 3.8 (Trace)

The trace of a matrix $\mathbf{A}$ is defined as

$$\mathrm{Tr}(\mathbf{A}) = \sum_k (\mathbf{A})_{kk}. \tag{3.176}$$

(a) Show that $\mathrm{Tr}(\mathbf{AB}) = \mathrm{Tr}(\mathbf{BA})$.

(b) Show that $\mathrm{Tr}(\mathbf{ABC}) = \mathrm{Tr}(\mathbf{BCA}) = \mathrm{Tr}(\mathbf{CAB})$.

(c) Show that $\mathrm{Tr}([\mathbf{A}, \mathbf{B}]) = 0$.

Exercise 3.9 (Matrix representation of 1D SHO)

Using the number states (or Fock states),[34] $\{|n\rangle\}_{n=0}^{\infty}$, i.e., the eigenstates of the 1D simple harmonic oscillator problem, as a basis, express the following operators as matrices: (i) $\hat{b}^{\dagger}$, (ii) $\hat{b}$, (iii) $\hat{n} = \hat{b}^{\dagger}\hat{b}$, (iv) $\hat{x}$, and (v) $\hat{p}_x$.

Exercise 3.10 (Pauli matrices)

The Pauli matrices σ_x, σ_y, and σ_z transform $|0\rangle$ and $|1\rangle$, given by Equation (3.155), in the following ways:

$$\sigma_x|0\rangle = |1\rangle, \ \sigma_x|1\rangle = |0\rangle, \tag{3.177}$$

$$\sigma_y|0\rangle = i|1\rangle, \ \sigma_y|1\rangle = -i|0\rangle, \tag{3.178}$$

$$\sigma_z|0\rangle = |0\rangle, \ \sigma_z|1\rangle = -|1\rangle. \tag{3.179}$$

(a) From these relations, derive the following:

$$\sigma_x = |0\rangle\langle 1| + |1\rangle\langle 0| = \begin{bmatrix} 0 & 1 \\ 1 & 0 \end{bmatrix}, \tag{3.180}$$

$$\sigma_y = -i|0\rangle\langle 1| + i|1\rangle\langle 0| = \begin{bmatrix} 0 & -i \\ i & 0 \end{bmatrix}, \tag{3.181}$$

$$\sigma_z = |0\rangle\langle 0| - |1\rangle\langle 1| = \begin{bmatrix} 1 & 0 \\ 0 & -1 \end{bmatrix}. \tag{3.182}$$

(b) Show that σ_x, σ_y, and σ_z are hermitian and unitary.

(c) The matrix σ_x is not diagonal. Find $\mathbf{U}$ such that $\mathbf{U}^{\dagger}\sigma_x\mathbf{U}$ is diagonal.

(d) Show that

$$\begin{bmatrix} 1 & 0 \\ 0 & 0 \end{bmatrix} = \frac{1}{2}(\mathbf{I}_{2D} + \sigma_z), \ \begin{bmatrix} 0 & 1 \\ 0 & 0 \end{bmatrix} = \frac{1}{2}(\sigma_x + i\sigma_y), \tag{3.183}$$

$$\begin{bmatrix} 0 & 0 \\ 1 & 0 \end{bmatrix} = \frac{1}{2}(\sigma_x - i\sigma_y), \ \begin{bmatrix} 0 & 0 \\ 0 & 1 \end{bmatrix} = \frac{1}{2}(\mathbf{I}_{2D} - \sigma_z). \tag{3.184}$$

Use these results to show that any 2×2 matrix can be expressed as a linear combination of σ_x, σ_y, σ_z, and $\mathbf{I}_{2D}$.

(e) Show the following property:

$$\sigma_i\sigma_k = \delta_{ik} + i\varepsilon_{ikl}\sigma_l, \tag{3.185}$$

where $i, k = 1, 2, 3$, $\sigma_1 := \sigma_x$, $\sigma_2 := \sigma_y$, $\sigma_3 := \sigma_z$, and

$$\varepsilon_{ikl} := \begin{cases} 1, & \text{if } (ikl) \text{ is } (123), (312), \text{ or } (231), \\ -1, & \text{if } (ikl) \text{ is } (132), (321), \text{ or } (213), \\ 0, & \text{if } i = k, k = l, \text{ or } l = i. \end{cases} \tag{3.186}$$

is the Levi–Civita symbol.

Exercise 3.11 (Measurement and uncertainty)

In a 3D Hilbert space with a certain basis, the Hamiltonian and an observable A are represented, respectively, by the following matrices:

$$\mathbf{H} = E_0 \begin{bmatrix} 1 & -1 & 0 \\ -1 & 1 & 0 \\ 0 & 0 & -1 \end{bmatrix}, \quad \mathbf{A} = a \begin{bmatrix} 0 & 4 & 0 \\ 4 & 0 & 1 \\ 0 & 1 & 0 \end{bmatrix}, \tag{3.187}$$

where E_0 and a are constants.

(a) Find the eigenvalues and normalized eigenvectors of $\mathbf{H}$. Denote the eigenvectors by $|\phi_1\rangle$, $|\phi_2\rangle$, and $|\phi_3\rangle$.

(b) Show that $|\phi_1\rangle$, $|\phi_2\rangle$, and $|\phi_3\rangle$ form an orthonormal and complete basis, i.e., $\sum_{j=1}^{3} |\phi_j\rangle\langle\phi_j| = \mathbf{I}_{3D}$ (where $\mathbf{I}_{3D}$ is the 3×3 unit matrix) and $\langle\phi_j|\phi_k\rangle = \delta_{jk}$.

(c) Find the eigenvalues and normalized eigenvectors of $\mathbf{A}$. Denote the eigenvectors by $|a_1\rangle$, $|a_2\rangle$, and $|a_3\rangle$.

(d) Show that $|a_1\rangle$, $|a_2\rangle$, and $|a_3\rangle$ form an orthonormal and complete basis, i.e., $\sum_{j=1}^{3} |a_j\rangle\langle a_j| = \mathbf{I}_{3D}$ (where $\mathbf{I}_{3D}$ is the 3×3 unit matrix) and $\langle a_j|a_k\rangle = \delta_{jk}$.

(e) Suppose that we measure the energy of this system. What are the possible values to be obtained and what are the probabilities of obtaining them?

(f) Suppose that we measure the energy and obtain a value of $-E_0$. If we measure A immediately afterwards, what values could be obtained and what are the probabilities of obtaining them?

(g) Calculate the expectation value $\langle A \rangle$ and the uncertainty $\Delta A = \sqrt{\langle A^2 \rangle - \langle A \rangle^2}$ in measuring A in the basis in (b).

4

Basics of Quantum Information Processing

Learning objectives:

- Understanding the definition and basic properties of qubits and their implementations.

- Appreciating the meaning and implications of quantum entanglement.

- Learning how to perform quantum gate operations to control qubits using unitary matrices.

- Becoming familiar with representative quantum circuits, quantum algorithms, and quantum hardware.

- Observing how quantum algorithms allow one to obtain a speedup in certain computational problems.

[1] Pronounced as 'kyoo·buht'.

THE FIELD OF QUANTUM COMPUTING is developing at a rapid pace, and one can expect paradigm-shifting advances in coming years. The goal of this chapter is for the reader to understand fully the language and basic concepts of quantum information science needed to engage in research and development in this very exciting field in the future. We will apply the mathematical machinery we have acquired so far to develop the quantum counterparts to the classical notions of bits, logic gates, circuits, and algorithms. We will also review some of the promising examples of quantum hardware for physically realizing quantum information processing.

4.1 Qubits

The fundamental building blocks in quantum information processing are called quantum bits, or qubits. A qubit[1] is a unit of computing information, and it is represented by a quantum state of an electron, a nucleus, or a photon. Let us first examine the basic properties of a single qubit, i.e., a single quantum system describable as a superposition state of two computational basis states, $|0\rangle$ and $|1\rangle$. We will then describe situations of multiple qubits, where we will encounter the nonintuitive concept of quantum entanglement, which is the essential enabler of some of the most exotic quantum technologies such as quantum teleportation.

4.1.1 Single Qubits

We saw in the last chapter that the eigenkets of a hermitian operator form a complete orthonormal basis set for the corresponding Hilbert space. Let $|0\rangle$ and $|1\rangle$ be two such eigenkets in a 2D Hilbert space. Because they form a complete basis, any arbitrary state ket $|\psi\rangle$ in this space can be written as

$$|\psi\rangle = \alpha|0\rangle + \beta|1\rangle, \tag{4.1}$$

where α and β are complex coefficients, and the normalization of $|\psi\rangle$ requires that $|\alpha|^2 + |\beta|^2 = 1$. Such a state $|\psi\rangle$ is said to be in a superposition of kets $|0\rangle$ and $|1\rangle$.

In quantum information science, any superposition of the basis vectors $|0\rangle$ and $|1\rangle$ represents a qubit. The set $\{|0\rangle, |1\rangle\}$ in this context is called the computational basis set. As we saw in Section 3.6, these basis kets can be represented by column vectors (or 2×1 matrices)

$$|0\rangle = \begin{bmatrix} 1 \\ 0 \end{bmatrix} \text{ and } |1\rangle = \begin{bmatrix} 0 \\ 1 \end{bmatrix} \tag{4.2}$$

whereas the bras $\langle 0|$ and $\langle 1|$ can be represented by row vectors (or 1×2 matrices)

$$\langle 0| = \begin{bmatrix} 1 & 0 \end{bmatrix} \text{ and } \langle 1| = \begin{bmatrix} 0 & 1 \end{bmatrix}. \tag{4.3}$$

The state of a single qubit can be geometrically thought of as a unit vector in a 2D complex vector space, which is conveniently visualized using a Bloch sphere; see Figure 4.1. Here, by association $\alpha := e^{i\gamma}\cos(\theta/2)$ and $\beta := e^{i(\gamma+\phi)}\sin(\theta/2)$ (which satisfy $|\alpha|^2 + |\beta|^2 = 1$), a qubit state can be written as

$$|\psi\rangle = e^{i\gamma}\left(\cos\frac{\theta}{2}|0\rangle + e^{i\phi}\sin\frac{\theta}{2}|1\rangle\right). \tag{4.4}$$

The resulting vector on the Bloch sphere is called a Bloch vector, and is a unit vector in this space whose orientation depends on the polar angle θ and the azimuthal angle ϕ. The overall phase (or global phase) γ has no physical consequences, but the phase ϕ is essential for quantum coherence to exist and it can cause, e.g., quantum interference phenomena.[2]

Each qubit state is *uniquely* represented by (θ, ϕ), i.e., a point on the Bloch sphere. For example, $(\theta, \phi) = (\frac{\pi}{2}, 0)$ represents $(|0\rangle + |1\rangle)/\sqrt{2}$, which is a 50%–50% superposition of $|0\rangle$ and $|1\rangle$. Note that the two special nonsuperposition states correspond to the "north pole" $|0\rangle$ ($\theta = 0$) and the "south pole" $|1\rangle$ ($\theta = \pi$). Since there is an infinite number of points on the Bloch sphere, there is an infinite number of possible superposition states. However, measurement of a qubit will only give either $|0\rangle$ or $|1\rangle$; the state collapses into one of them after measurement (see Section 3.5). The Bloch sphere is convenient particularly because it allows us to visualize the effects of gates (i.e., unitary operations) on a single qubit; see Section 4.3.

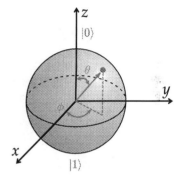

Figure 4.1 The Bloch sphere, representing a single qubit. © Deyin Kong (Rice University).

[2] See, e.g., Remark 2.1.

4.1.2   Multiple Qubits

While single qubits provide a quantum coherent basis on which to build quantum-engineered devices, they alone cannot be used for performing computation. In order to implement quantum algorithms,[3] one needs to create an assembly of qubits, or multiple qubits, and realize a special type of interactions, i.e., quantum entanglement, among the individual qubits.

[3] See Section 4.6.

Let us start with the case of two qubits – Qubit 1 and Qubit 2. In general, the state of this two-qubit system can be represented by product states such as $|0\rangle_1|0\rangle_2$ (or more commonly written simply as $|00\rangle$). This particular product state represents a two-qubit state in which Qubit 1 is in state $|0\rangle_1$ and Qubit 2 is in state $|0\rangle_2$. The subscripts "1" and "2" are often omitted. There are three other such well-defined product states – $|01\rangle$, $|10\rangle$, and $|11\rangle$ – and together, the computational basis set $\{|00\rangle, |01\rangle, |10\rangle, |11\rangle\}$ can describe all states in this two-qubit system.[4] Namely, a general state in this four-dimensional (4D) Hilbert space can be written as

[4] The binary numbers that these states represent can be converted to decimal numbers as $00_2 = 0 \times 2^1 + 0 \times 2^0 = 0$, $01_2 = 0 \times 2^1 + 1 \times 2^0 = 1$, $10_2 = 1 \times 2^1 + 0 \times 2^0 = 2$, and $11_2 = 1 \times 2^1 + 1 \times 2^0 = 3$.

$$|\psi\rangle = \alpha_{00}|00\rangle + \alpha_{01}|01\rangle + \alpha_{10}|10\rangle + \alpha_{11}|11\rangle, \tag{4.5}$$

where the four expansion coefficients must satisfy the usual relation

$$|\alpha_{00}|^2 + |\alpha_{01}|^2 + |\alpha_{10}|^2 + |\alpha_{11}|^2 = 1 \tag{4.6}$$

for $|\psi\rangle$ to be normalized.

Representing product states by matrices requires the use of the tensor product,[5] denoted by the symbol $\otimes$, which creates a third, larger Hilbert space, H_3, from two Hilbert spaces, H_1 and H_2. Namely,

[5] It is also known as the Kronecker product.

$$|\psi_1\rangle \in H_1, |\psi_2\rangle \in H_2 \;\Rightarrow\; |\psi_1\rangle \otimes |\psi_2\rangle \in H_3. \tag{4.7}$$

Formally, if $\mathbf{A}$ is an $m \times n$ matrix and $\mathbf{B}$ is a $p \times q$ matrix, then the tensor product $\mathbf{A} \otimes \mathbf{B}$ is the following $pm \times qn$ block matrix:

$$\mathbf{A} \otimes \mathbf{B} = \begin{bmatrix} a_{11}\mathbf{B} & a_{12}\mathbf{B} & \dots & a_{1n}\mathbf{B} \\ a_{21}\mathbf{B} & a_{22}\mathbf{B} & \dots & a_{2n}\mathbf{B} \\ \vdots & \vdots & & \vdots \\ a_{m1}\mathbf{B} & a_{m2}\mathbf{B} & \dots & a_{mn}\mathbf{B} \end{bmatrix}. \tag{4.8}$$

For example, the matrix representation of the product state $|01\rangle$ is

$$|01\rangle = |0\rangle \otimes |1\rangle = \begin{bmatrix} 1 \times \begin{bmatrix} 0 \\ 1 \end{bmatrix} \\ 0 \times \begin{bmatrix} 0 \\ 1 \end{bmatrix} \end{bmatrix} = \begin{bmatrix} 0 \\ 1 \\ 0 \\ 0 \end{bmatrix}. \tag{4.9}$$

Note that $|0\rangle \otimes |1\rangle$ is a $(2 \times 2) \times (1 \times 1) = 4 \times 1$ matrix because both $|0\rangle$ and $|1\rangle$ are 2×1 matrices. When the two matrices in the product are identical, the symbol $^{\otimes 2}$ is used. For example,[6]

$$|00\rangle = |0\rangle \otimes |0\rangle = |0\rangle^{\otimes 2} = \begin{bmatrix} 1 \times \begin{bmatrix} 1 \\ 0 \end{bmatrix} \\ 0 \times \begin{bmatrix} 1 \\ 0 \end{bmatrix} \end{bmatrix} = \begin{bmatrix} 1 \\ 0 \\ 0 \\ 0 \end{bmatrix}. \qquad (4.10)$$

Similarly, the state of a three-qubit system is represented by product states such as $|010\rangle$, and the dimension of the Hilbert space is $2^3 = 8$. In general, a state in an N-qubit system is a ket vector in a 2^N-dimensional Hilbert space.[7]

4.2 Quantum Entanglement

Let us consider a two-qubit quantum system for which a general state is described by the superposition state given by Equation (4.5). The probabilistic interpretation of the expansion coefficients α_{ij} in this equation remains the same as in the single-qubit case. For example, $|\alpha_{01}|^2$ represents the probability that Qubit 1 is found to be in the state $|0\rangle$ and Qubit 2 is found to be in the state $|1\rangle$.

Suppose that we perform a quantum measurement on the first qubit and find it to be in the state $|0\rangle$. This occurs with probability $|\alpha_{00}|^2 + |\alpha_{01}|^2$. Obviously, this knowledge itself does not tell us anything about the second qubit. What we know is that, after the measurement, the system is (due to wavefunction collapse) in the state

$$|\psi'\rangle = \frac{\alpha_{00}|00\rangle + \alpha_{01}|01\rangle}{\sqrt{|\alpha_{00}|^2 + |\alpha_{01}|^2}}. \qquad (4.13)$$

If we further perform a quantum measurement on the second qubit, we find it to be either in $|0\rangle$ with probability $|\alpha_{00}|^2 / (|\alpha_{00}|^2 + |\alpha_{01}|^2)$ or in $|1\rangle$ with probability $|\alpha_{01}|^2 / (|\alpha_{00}|^2 + |\alpha_{01}|^2)$.

Now let us consider the following special state, called a Bell state:

$$|\phi^+\rangle = \frac{|00\rangle + |11\rangle}{\sqrt{2}}. \qquad (4.14)$$

This corresponds to the general wavefunction given by Equation (4.5) when $\alpha_{00} = \alpha_{11} = \frac{1}{\sqrt{2}}$ and $\alpha_{01} = \alpha_{10} = 0$. Suppose now that we measure the first qubit and find it to be in $|0\rangle$. What is amazing is that, in this case, we already know that the second qubit is also in $|0\rangle$ *without performing a measurement on the second qubit!* This is true even when

[6] The other two basis vectors are expressed as

$$|10\rangle = \begin{bmatrix} 0 \\ 0 \\ 1 \\ 0 \end{bmatrix}, \quad |11\rangle = \begin{bmatrix} 0 \\ 0 \\ 0 \\ 1 \end{bmatrix}.$$

[7] There are two other important relations that are often useful (for, e.g., Exercise 4.3):

$$\langle f_1 \otimes f_2 | g_1 \otimes g_2 \rangle = \langle f_1 | g_1 \rangle_{H_1} \langle f_2 | g_2 \rangle_{H_2} \qquad (4.11)$$

$$(\hat{A}_1 \otimes \hat{A}_2)(f_1 \otimes f_2) = (\hat{A}_1 f_1) \otimes (\hat{A}_2 f_2) \qquad (4.12)$$

Here, the states $|f_1\rangle$ and $|g_1\rangle$ and operator $\hat{A}_1$ belong to Hilbert space H_1, the states $|f_2\rangle$ and $|g_2\rangle$ and operator $\hat{A}_2$ belong to Hilbert space H_2, and their tensor products produce a larger Hilbert space H_3.

the two qubits are physically far away from each other.[8] Similarly, if our measurement finds the first qubit to be in $|1\rangle$, we automatically know that, if we measure the second qubit, it will also be found to be in $|1\rangle$ *with 100% probability*. Hence, measurement outcomes are correlated. This nonclassical correlation is called quantum entanglement, and the state $|\phi^+\rangle$ is said to be an entangled state.

There are actually four Bell states, given by

$$|\psi^+\rangle = \frac{|01\rangle + |10\rangle}{\sqrt{2}}, \tag{4.15}$$

$$|\psi^-\rangle = \frac{|01\rangle - |10\rangle}{\sqrt{2}}, \tag{4.16}$$

$$|\phi^+\rangle = \frac{|00\rangle + |11\rangle}{\sqrt{2}}, \tag{4.17}$$

$$|\phi^-\rangle = \frac{|00\rangle - |11\rangle}{\sqrt{2}}, \tag{4.18}$$

and it can be easily shown that they all represent quantum entangled states. For three-qubit systems, the following state (known as the Greenberger–Horne–Zeilinger state)

$$|\text{GHZ}\rangle = \frac{|000\rangle + |111\rangle}{\sqrt{2}} \tag{4.19}$$

represents an entangled state. As soon as one of the three qubits is measured, the other two qubits will each assume a well-defined value.

Note that, in all these entangled states, an individual qubit does not carry any well-defined information on its own; all the information is encoded in their *joint properties*. For example, let us consider a two-qubit system consisting of two spin-$\frac{1}{2}$ particles with single-qubit basis kets $|\uparrow\rangle$ (spin "up") and $|\downarrow\rangle$ (spin "down"). If the system is in one of the Bell states $|\psi^+\rangle = \frac{1}{\sqrt{2}}(|01\rangle + |10\rangle) = \frac{1}{\sqrt{2}}(|\uparrow\downarrow\rangle + |\downarrow\uparrow\rangle)$, we have no information on the spin orientations of the individual particles, but we know that the total spin of this two-qubit system is zero since the two spins are always antiparallel.

Mathematically, $|\psi\rangle \in H_1 \otimes H_2$ is said to be disentangled or separable if there exist states $|\psi_1\rangle \in H_1$ and $|\psi_2\rangle \in H_2$ such that $|\psi\rangle = |\psi_1\rangle \otimes |\psi_2\rangle$.[9] In other words, a state $|\psi\rangle \in H_1 \otimes H_2$ is disentangled when it is a product state. A corollary is that an entangled state cannot be a product state. An entangled state is a *superposition of product states*. For example, the Bell state $|\psi^+\rangle = \frac{1}{\sqrt{2}}(|01\rangle + |10\rangle)$ is a superposition of the two product states $|01\rangle$ and $|10\rangle$. For a general two-qubit state given by Equation (4.5), it can be shown that it is an entangled state if $\alpha_{00}\alpha_{11} \neq \alpha_{01}\alpha_{10}$.[10]

4.3 Quantum Gates

Before delving into quantum gates, let us quickly review classical gates, since logic gates (or simply gates) are important components in any digital system. A gate is a physical device that implements a Boolean function, taking one or more binary inputs and producing a single binary output. For example, one of the simplest one-input–one-output classical gates is the NOT gate (or also known as the inverter), whose truth table is shown in Table 4.1. Here, "1" and "0" represent Boolean data, *true* and *false*, respectively.

INPUT	OUTPUT
0	1
1	0

Table 4.1 Truth table for the classical NOT gate (or the inverter).

Furthermore, Table 4.2 provides the truth table for six common examples of classical two-input–one-output logic gates. Here, the "A" and "B" values are the two input values, while the six columns on the right list the output values for the six gates, AND, NAND, OR, NOR, XOR, and XNOR, respectively. The output of the AND gate is 1 only when both input A and input B are 1. The NAND gate produces an output that is 1 only if both inputs are 1, so its output is always opposite to that of the AND gate. The OR gate gives an output of 1 when any of its inputs are 1, otherwise 0. The NOR gate is the negation of the OR gate. Finally, the XOR gate (or the exclusive OR gate) gives a 1 output if one, and only one, of the inputs is true. If both inputs are 0 or both are 1, a 0 output results. It can be shown that that NOR gates alone (or alternatively NAND gates alone) can be used to reproduce the functions of all the other logic gates.

INPUT		OUTPUT					
A	B	AND	NAND	OR	NOR	XOR	XNOR
0	0	0	1	0	1	0	1
0	1	0	1	1	0	1	0
1	0	0	1	1	0	1	0
1	1	1	0	1	0	0	1

Table 4.2 Truth table for the classical two-input–one-output logic gates. It can be shown that NOR gates alone (or alternatively NAND gates alone) can be used to reproduce the functions of all the other logic gates.

4.3.1 Single-Qubit Gates

In this subsection we describe quantum gates that operate on single qubits, i.e., gates that take a single-qubit state, $|\psi_{\text{in}}\rangle$, as an input and produce another single-qubit state, $|\psi_{\text{out}}\rangle$, as an output. Specifically, we will use the NOT gate and the Hadamard gate to learn the basic properties of single-qubit gates. Two other important single-qubit gates, the Z gate and the phase shifter, are described in Exercise 4.5.

Similarly to the classical NOT gate (Table 4.1), the quantum NOT gate changes $|0\rangle$ into $|1\rangle$ and changes $|1\rangle$ into $|0\rangle$. Namely, the linear operator that represents the NOT gate, $\hat{U}_{\text{NOT}}$, can be defined through its actions on $|0\rangle$ and $|1\rangle$:

$$\hat{U}_{\mathrm{NOT}}|0\rangle = |1\rangle, \qquad (4.20)$$

$$\hat{U}_{\mathrm{NOT}}|1\rangle = |0\rangle. \qquad (4.21)$$

[11] See Equation (3.103).

Further, by multiplying the identity operator[11] $\hat{1} = |0\rangle\langle 0| + |1\rangle\langle 1|$ on $\hat{U}_{\mathrm{NOT}}$ from the right, we get

$$\hat{U}_{\mathrm{NOT}} = \hat{U}_{\mathrm{NOT}}(|0\rangle\langle 0| + |1\rangle\langle 1|) = |1\rangle\langle 0| + |0\rangle\langle 1|. \qquad (4.22)$$

Recalling Equations (4.2) and (4.3), we can then express $\hat{U}_{\mathrm{NOT}}$ by the following matrix:

$$\mathbf{U}_{\mathrm{NOT}} = \begin{bmatrix} 1 \\ 0 \end{bmatrix} \begin{bmatrix} 0 & 1 \end{bmatrix} + \begin{bmatrix} 0 \\ 1 \end{bmatrix} \begin{bmatrix} 1 & 0 \end{bmatrix} = \begin{bmatrix} 0 & 1 \\ 1 & 0 \end{bmatrix} = \sigma_x, \qquad (4.23)$$

i.e., the quantum NOT gate is represented by the Pauli matrix σ_x.

Accordingly, in a quantum circuit (see Section 4.4 for more details), a NOT gate appears as in Figure 4.2. Here, an input single qubit, $|\psi_{\mathrm{in}}\rangle$, comes into this gate from the left, and an output single qubit, $|\psi_{\mathrm{out}}\rangle$, goes out of this gate from the right. Note that since $\hat{U}_{\mathrm{NOT}}$ is a linear operator, it converts a superposition state into another superposition state. This can be readily seen as follows:

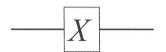

Figure 4.2 The quantum NOT gate. This gate changes an input single qubit $|\psi_{\mathrm{in}}\rangle = \alpha|0\rangle + \beta|1\rangle$ into an output single qubit $|\psi_{\mathrm{out}}\rangle = \alpha|1\rangle + \beta|0\rangle$.

$$\hat{U}_{\mathrm{NOT}}(\alpha|0\rangle + \beta|1\rangle) = \alpha\hat{U}_{\mathrm{NOT}}|0\rangle + \beta\hat{U}_{\mathrm{NOT}}|1\rangle$$
$$= \alpha|1\rangle + \beta|0\rangle. \qquad (4.24)$$

This property, known as quantum parallelism, can lead to a massive acceleration of computation.[12]

[12] See Section 4.6.1.

Another important property of $\hat{U}_{\mathrm{NOT}}$ is that it is *unitary*, i.e.,

$$\hat{U}_{\mathrm{NOT}}^{\dagger}\hat{U}_{\mathrm{NOT}} = \hat{1}. \qquad (4.25)$$

This can be seen immediately from the properties

$$\sigma_x^{\dagger} = \sigma_x^{-1} = \sigma_x \qquad (4.26)$$

of the Pauli matrix σ_x. The fact that $\hat{U}_{\mathrm{NOT}}$ has an inverse means that it is *reversible*. This is an important property that distinguishes quantum gates from classical gates, because, as is evident from Table 4.2, many classical gates are irreversible. For example, in the classical AND gate, an output value (either 0 or 1) cannot uniquely tell us the two original input values. In the case of the quantum NOT gate, however, if the output qubit is $|1\rangle$, the input must be $|0\rangle$, and if the output is $|0\rangle$, then the input must be $|1\rangle$.

Interestingly, *any* unitary operator, $\hat{U}$, specifies a valid quantum gate, and any such gate operator changes a state ket vector in the 2D

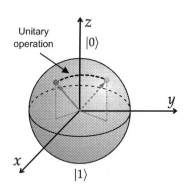

Figure 4.3 A unitary operation on a single qubit can be visualized as a rotation of a unit vector whose tip is on the Bloch sphere. © Deyin Kong (Rice University).

Hilbert space into another ket vector *while preserving its length*. Since the original ket $|\psi_{\text{in}}\rangle = \alpha|0\rangle + \beta|1\rangle$ is a unit vector, $|\alpha|^2 + |\beta|^2 = 1$. Hence, the unitarity of $\hat{U}$ ensures that $|\psi_{\text{out}}\rangle = \hat{U}|\psi_{\text{in}}\rangle = \alpha'|0\rangle + \beta'|1\rangle$ is a unit vector, i.e., $|\alpha'|^2 + |\beta'|^2 = 1$. From this point of view, any gate operation on a single qubit can be considered to be a "rotation" of the ket vector in Hilbert space whose tip is always on the Bloch sphere, as shown in Figure 4.3. Mathematically, in fact, any single-qubit operation can be decomposed as

$$\mathbf{U} = e^{i\gamma}\begin{bmatrix} e^{-i\phi/2} & 0 \\ 0 & e^{i\phi/2} \end{bmatrix}\begin{bmatrix} \cos(\theta/2) & -\sin(\theta/2) \\ \sin(\theta/2) & \cos(\theta/2) \end{bmatrix}. \tag{4.27}$$

Physically, a quantum gate is a system with a certain Hamiltonian, $\hat{H}$, and a unitary operation corresponds to the time evolution operator,[13] $\hat{U}(t) = e^{-i\hat{H}t/\hbar}$, which is naturally unitary. While the system is evolving in time under the action of the time evolution operator, the state vector evolves as $|\psi(t)\rangle = \hat{U}(t)|0\rangle$, with its norm always preserved as 1, i.e.,

[13] See Equation (2.81).

$$\langle\psi(t)|\psi(t)\rangle = \hat{U}^{\dagger}\hat{U}\langle\psi(t)|\psi(t)\rangle = e^{i\hat{H}t/\hbar}e^{-i\hat{H}t/\hbar}\langle\psi(0)|\psi(0)\rangle$$
$$= \langle\psi(0)|\psi(0)\rangle = 1. \tag{4.28}$$

Another important single-qubit gate is the Hadamard gate, $\hat{U}_{\text{H}}$, which transforms $|0\rangle$ and $|1\rangle$ as follows:

$$|0\rangle \rightarrow \frac{1}{\sqrt{2}}(|0\rangle + |1\rangle), \tag{4.29}$$

$$|1\rangle \rightarrow \frac{1}{\sqrt{2}}(|0\rangle - |1\rangle). \tag{4.30}$$

Therefore, its matrix representation, $\mathbf{H}$, using the basis given by Equations (4.2) and (4.3), can be obtained as

$$\mathbf{H} = \mathbf{H}(|0\rangle\langle0| + |1\rangle\langle1|) = \frac{1}{\sqrt{2}}\left\{(|0\rangle + |1\rangle)\langle0| + (|0\rangle - |1\rangle)\langle1|\right\}$$

$$= \frac{1}{\sqrt{2}}(|0\rangle\langle0| + |0\rangle\langle1| + |1\rangle\langle0| - |1\rangle\langle1|)$$

$$= \frac{1}{\sqrt{2}}\left\{\begin{bmatrix}1\\0\end{bmatrix}\begin{bmatrix}1 & 0\end{bmatrix} + \begin{bmatrix}1\\0\end{bmatrix}\begin{bmatrix}0 & 1\end{bmatrix} + \begin{bmatrix}0\\1\end{bmatrix}\begin{bmatrix}1 & 0\end{bmatrix} - \begin{bmatrix}0\\1\end{bmatrix}\begin{bmatrix}0 & 1\end{bmatrix}\right\}$$

$$= \frac{1}{\sqrt{2}}\begin{bmatrix}1 & 1\\1 & -1\end{bmatrix}. \tag{4.31}$$

One can see that $\mathbf{H}$ is hermitian ($\mathbf{H}^{\dagger} = \mathbf{H}$) and unitary ($\mathbf{H}^{\dagger} = \mathbf{H}^{-1}$), and thus, $\mathbf{H}^2 = \mathbf{I}_{\text{2D}}$. In a quantum circuit, the Hadamard gate appears as in Figure 4.4.

Figure 4.4 The Hadamard gate changes an input single qubit $|\psi_{\text{in}}\rangle = \alpha|0\rangle + \beta|1\rangle$ into an output single qubit

$$|\psi_{\text{out}}\rangle = \frac{\alpha}{\sqrt{2}}(|0\rangle + |1\rangle) + \frac{\beta}{\sqrt{2}}(|0\rangle - |1\rangle)$$

$$= \frac{1}{\sqrt{2}}\left\{(\alpha + \beta)|0\rangle + (\alpha - \beta)|1\rangle\right\}.$$

4.3.2   Multiple-Qubit Gates

Quite surprisingly, when we go from a single-qubit situation to a multiple-qubit situation, there is only one new quantum gate that we need to learn. This crucially important two-qubit gate is called the controlled-NOT gate (also known as the C-NOT or CNOT gate), and it can be shown that any multiple-qubit logic gate can be composed from CNOT and single-qubit gates.

Table 4.3 Truth table for the controlled NOT (or CNOT) gate. $|A\rangle_c$ is the "control" qubit, whereas $|B\rangle_t$ is the "target" qubit.

| INPUT		OUTPUT	
$|A\rangle_c$	$|B\rangle_t$	$|A\rangle_c$	$|B\rangle_t$
$|0\rangle$	$|0\rangle$	$|0\rangle$	$|0\rangle$
$|0\rangle$	$|1\rangle$	$|0\rangle$	$|1\rangle$
$|1\rangle$	$|0\rangle$	$|1\rangle$	$|1\rangle$
$|1\rangle$	$|1\rangle$	$|1\rangle$	$|0\rangle$

Table 4.3 presents the truth table for the CNOT gate, which describes how this gate works. There are two quantum states, or qubits, $|A\rangle_c$ and $|B\rangle_t$, and the input is their product $|\psi_{in}\rangle = |A\rangle_c|B\rangle_t = |AB\rangle$. Note that the qubit $|A\rangle_c$ remains unchanged while going through this gate. On the other hand, the qubit $|B\rangle_t$ does change, *depending on the value of* $|A\rangle_c$. Specifically, the value of $|B\rangle_t$ is *negated* (i.e., $|0\rangle \to |1\rangle$ or $|1\rangle \to |0\rangle$) whenever $|A\rangle_c = |1\rangle$ (i.e., whenever it is *true*); the value of $|B\rangle_t$ remains unchanged when $|A\rangle_c = |0\rangle$ (i.e., when it is *false*). This is why $|A\rangle_c$ is called the control qubit, whereas $|B\rangle_t$ is called the target qubit.

Figure 4.5 shows the CNOT gate as it appears as a component in a quantum circuit. Here, there are two single qubits that enter this gate from the left, and there are two single qubits that go out of this gate from the right. The control qubit, $|A\rangle_c$, evolves in time along the top line while the target qubit, $|B\rangle_t$, evolves in time along the bottom line, again from the left to the right. As shown in the truth table (Table 4.3), $|A\rangle_c$ remains unchanged while $|B\rangle_t$ changes its value only when $|A\rangle_c = |1\rangle$.

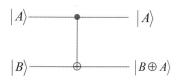

Figure 4.5 The CNOT gate. This gate changes an input qubit $|\psi_{in}\rangle = |A\rangle_c|B\rangle_t$ into an output qubit $|\psi_{out}\rangle = |A\rangle_c|B \oplus A\rangle_t$.

Note that the symbol $\oplus$ means "addition modulo 2," i.e., $0 \oplus 0 = 0$, $0 \oplus 1 = 1, 1 \oplus 0 = 1$, and $1 \oplus 1 = 0$. From the first and fourth relations here, one can conclude that $A \oplus A = 0$ for any $A \in \{0,1\}$. Note also that $A \oplus B = A$ XOR B. Furthermore, the following commutative and associative laws hold, as in normal addition: $A \oplus B = B \oplus A$ and $(A \oplus B) \oplus C = A \oplus (B \oplus C) = A \oplus B \oplus C$ for $A, B, C \in \{0,1\}$.

Let us now consider how to represent the CNOT operator, $\hat{U}_{CNOT}$, as a matrix. By definition (given by Table 4.3), this operator transforms the basis vectors of the 4D Hilbert space corresponding to this two-qubit system as follows: $\hat{U}_{CNOT}|00\rangle = |00\rangle$, $\hat{U}_{CNOT}|01\rangle = |01\rangle$,

$\hat{U}_{CNOT}|10\rangle = |11\rangle$, and $\hat{U}_{CNOT}|11\rangle = |10\rangle$. Thus, we can obtain the matrix as

$$\mathbf{U}_{CNOT} = \mathbf{U}_{CNOT}(|00\rangle\langle 00| + |01\rangle\langle 01| + |10\rangle\langle 10| + |11\rangle\langle 11|)$$
$$= |00\rangle\langle 00| + |01\rangle\langle 01| + |11\rangle\langle 10| + |10\rangle\langle 11|$$
$$= \begin{bmatrix} 1 & 0 & 0 & 0 \\ 0 & 1 & 0 & 0 \\ 0 & 0 & 0 & 1 \\ 0 & 0 & 1 & 0 \end{bmatrix}. \tag{4.32}$$

Note that $\mathbf{U}_{CNOT}$ is unitary and thus it is *reversible*, which distinguishes this quantum two-qubit gate from classical two-input gates (see Table 4.2), as in the case of single-qubit gates.

4.4 Quantum Circuits

We have already seen some of the simplest quantum circuits (see, e.g., Figures 4.2, 4.4, and 4.5). Time progresses from left to right in any quantum circuit, describing how qubits evolve. Each solid line carries a single qubit $|\psi\rangle = \alpha|0\rangle + \beta|1\rangle$ [Figure 4.6(a)], while n qubits $|\psi_1 \psi_2 \dots \psi_n\rangle$ can be carried by a wire, as shown in Figure 4.6(b). When there is a gate on a solid line as in Figures 4.2, 4.4, and 4.5, a qubit enters the gate from the left, evolves under the action of the unitary operator of the gate, and exits onto the line on the right-hand side of the gate. A double solid line carries a *classical* bit (0 or 1) [Figure 4.6(c)]. When a measurement is performed, a single qubit is destroyed, producing a classical bit [Figure 4.6(d)].

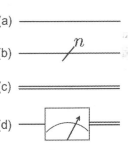

Figure 4.6 Major components in quantum circuits. (a) Wire carrying a single qubit. (b) Wire carrying n qubits. (c) Wire carrying a classical bit. (d) Quantum measurement, converting a single qubit (on the left) into a classical bit (on the right).

Example 4.1 The swap circuit

Let us consider the circuit shown in Figure 4.7. This is a two-qubit circuit, as one can tell from the two solid horizontal lines. We can follow the evolution of the system wavefunction and then understand the relationship between the initial and final states. By letting $|\psi_1\rangle = |A, B\rangle$ be the initial state, we see that it first goes through a CNOT gate to become $|\psi_2\rangle = |A, A \oplus B\rangle$. Next, it again goes through a CNOT gate, but the roles of control and target qubits are reversed compared with the first CNOT gate. Thus,

$$|\psi_3\rangle = |A \oplus (A \oplus B), A \oplus B\rangle = |B, A \oplus B\rangle. \tag{4.33}$$

Here, we have used the associative law, $A \oplus (A \oplus B) = (A \oplus A) \oplus B$, together with the fact that $A \oplus A = 0$ for any $A \in \{0,1\}$. Finally, the qubit goes through another CNOT gate to become

$$|\psi_4\rangle = |B, (A \oplus B) \oplus B\rangle = |B, A\rangle. \tag{4.34}$$

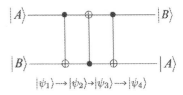

Figure 4.7 The swap circuit.

INPUT		OUTPUT	
$|A\rangle$	$|B\rangle$	$|A\rangle$	$|B\rangle$
$|0\rangle$	$|0\rangle$	$|0\rangle$	$|0\rangle$
$|0\rangle$	$|1\rangle$	$|1\rangle$	$|0\rangle$
$|1\rangle$	$|0\rangle$	$|0\rangle$	$|1\rangle$
$|1\rangle$	$|1\rangle$	$|1\rangle$	$|1\rangle$

Table 4.4 Truth table for the swap circuit.

Therefore, we can construct the corresponding truth table, Table 4.4. The table demonstrates that this circuit swaps the values of the two input qubits.

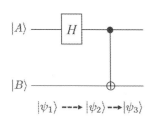

Figure 4.8 The quantum circuit that produces Bell states.

INPUT		OUTPUT
$\lvert A \rangle$	$\lvert B \rangle$	$\lvert \psi_3 \rangle$
$\lvert 0 \rangle$	$\lvert 0 \rangle$	$\frac{1}{\sqrt{2}}(\lvert 00 \rangle + \lvert 11 \rangle) = \lvert \phi^+ \rangle$
$\lvert 0 \rangle$	$\lvert 1 \rangle$	$\frac{1}{\sqrt{2}}(\lvert 01 \rangle + \lvert 10 \rangle) = \lvert \psi^+ \rangle$
$\lvert 1 \rangle$	$\lvert 0 \rangle$	$\frac{1}{\sqrt{2}}(\lvert 00 \rangle - \lvert 11 \rangle) = \lvert \phi^- \rangle$
$\lvert 1 \rangle$	$\lvert 1 \rangle$	$\frac{1}{\sqrt{2}}(\lvert 01 \rangle - \lvert 10 \rangle) = \lvert \psi^- \rangle$

Table 4.5 Truth table for the Bell state generator.

Example 4.2 The Bell state generator

Next, let us consider the two-qubit circuit shown in Figure 4.8, known as the Bell state generator. In the first stage, the qubit $\lvert A \rangle$ is modified by the Hadamard gate, and in the second stage, the two-qubit state goes through a CNOT gate. Depending on the input qubits, this circuit can generate the Bell states, i.e., Equations (4.15), (4.16), (4.17), and (4.18). For example, by inputting $\lvert \psi_1 \rangle = \lvert AB \rangle = \lvert 00 \rangle$, we can see how the state evolves, as follows:

$$\lvert \psi_2 \rangle = \frac{1}{\sqrt{2}}(\lvert 0 \rangle + \lvert 1 \rangle)\lvert 0 \rangle$$

$$= \frac{1}{\sqrt{2}}(\lvert 00 \rangle + \lvert 10 \rangle), \tag{4.35}$$

$$\lvert \psi_3 \rangle = \frac{1}{\sqrt{2}}(\lvert 00 \rangle + \lvert 11 \rangle). \tag{4.36}$$

One can see immediately that $\lvert \psi_3 \rangle = \lvert \phi^+ \rangle$, i.e., Equation (4.17). Similarly, it can be shown that $\lvert \psi_1 \rangle = \lvert 01 \rangle$ generates $\lvert \psi_3 \rangle = \lvert \psi^+ \rangle$, $\lvert \psi_1 \rangle = \lvert 10 \rangle$ generates $\lvert \psi_3 \rangle = \lvert \phi^- \rangle$, and $\lvert \psi_1 \rangle = \lvert 11 \rangle$ generates $\lvert \psi_3 \rangle = \lvert \psi^- \rangle$. Table 4.5 is the truth table for this circuit.

4.5 Quantum Teleportation

Figure 4.9 Schematic diagram of the quantum teleportation protocol proposed by C. H. Bennett *et al.* [*Physical Review Letters* **70**, 1895 (1993)]. © Deyin Kong (Rice University).

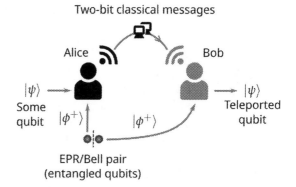

By combining some of the gates we have learned about so far, together with the concepts of entanglement and quantum measurement, we can construct a quantum circuit that performs the quantum teleportation protocol proposed by Bennett and coworkers in 1993; see Figures 4.9 and 4.10. This circuit enables the transfer of a quantum state between two locations (say, locations A and B). More specifically, an exact replica of an original state $|\psi\rangle$ at location A can be created at location B. The original state $|\psi\rangle$ is destroyed during this process and, thus, this is not a cloning process. Note also that the actual "object" (i.e., the qubit) does not traverse the distance between the two locations. The object is measured at A to extract sufficient information to recreate the original, the information is transmitted, and an exact replica is assembled at B out of locally available material based on the information. In this sense, FAX transmission is distinct from teleportation, since the original remains intact and the copy is not exact in FAX transmission.

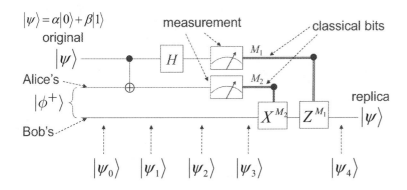

Figure 4.10 A quantum circuit for accomplishing the quantum teleportation task depicted in Figure 4.9. The original state $|\psi\rangle = \alpha|0\rangle + \beta|1\rangle$ initially exists in Alice's location (location A), and its exact replica is created at Bob's location (location B). $|\phi^+\rangle$ is one of the Bell states, Equation (4.17). M_1 and M_2 are classical bits, 0 or 1.

Let us go through this circuit step by step. The first step is for Alice (at location A) to create a Bell pair, say, $|\phi^+\rangle = (|00\rangle + |11\rangle)/\sqrt{2}$, which is one of the Bell states [Equation (4.17)]. Alice keeps one of the entangled qubits and gives the other to Bob. In the circuit, there are three lines, representing three qubits. In the initial state $|\psi_0\rangle$, the first qubit corresponds to the original state $|\psi\rangle$, while the second and third lines correspond to the Bell pair. Thus,

$$|\psi_0\rangle = |\psi\rangle|\phi^+\rangle = (\alpha|0\rangle + \beta|1\rangle)\frac{|00\rangle + |11\rangle}{\sqrt{2}}$$
$$= \frac{1}{\sqrt{2}}\{\alpha|0\rangle(|00\rangle + |11\rangle) + \beta|1\rangle(|00\rangle + |11\rangle)\}. \qquad (4.37)$$

In the next step of the circuit, $|\psi_0\rangle$ goes through a CNOT gate to become $|\psi_1\rangle$. In this process, the first qubit, the control qubit, does not change while the second qubit changes when the control gate is $|1\rangle$ ("true"). Therefore, we can write

$$|\psi_1\rangle = \frac{1}{\sqrt{2}}\{\alpha|0\rangle(|00\rangle + |11\rangle) + \beta|1\rangle(|10\rangle + |01\rangle)\}. \qquad (4.38)$$

Then the first qubit goes through a Hadamard gate, so that

$$|\psi_2\rangle = \frac{1}{2}\{\alpha(|0\rangle + |1\rangle)(|00\rangle + |11\rangle) + \beta(|0\rangle - |1\rangle)(|10\rangle + |01\rangle)\}$$

$$= \frac{1}{2}[|00\rangle(\alpha|0\rangle + \beta|1\rangle) + |01\rangle(\alpha|1\rangle + \beta|0\rangle)$$

$$+ |10\rangle(\alpha|0\rangle - \beta|1\rangle) + |11\rangle(\alpha|1\rangle - \beta|0\rangle)]. \quad (4.39)$$

Note that there are four terms in the final expression above, corresponding to the four possibilities when the first two qubits are measured by Alice in the next step. If the measurement results are $(M_1, M_2) = (0,0)$, then $|\psi_3\rangle = \alpha|1\rangle + \beta|0\rangle$, which is already the original state. In this case, Alice tells Bob not to do anything since Bob already has a replica of the original state. This corresponds to two successive operations by the identity operator, i.e.,

$$|\psi_4\rangle = (\sigma_z)^0(\sigma_x)^0|\psi_3\rangle = \hat{1}\hat{1}|\psi_3\rangle = \alpha|0\rangle + \beta|1\rangle = |\psi\rangle. \quad (4.40)$$

If $(M_1, M_2) = (0,1)$, on the other hand, $|\psi_3\rangle = \alpha|1\rangle + \beta|0\rangle$. Hence, Alice tells Bob to apply the X gate once to recover $|\psi\rangle$:

$$|\psi_4\rangle = (\sigma_z)^0(\sigma_x)^1|\psi_3\rangle = \sigma_x(\alpha|1\rangle + \beta|0\rangle) = \alpha|0\rangle + \beta|1\rangle = |\psi\rangle. \quad (4.41)$$

In a similar manner, Bob can create a replica by operating σ_z and $\sigma_x\sigma_z$ on his qubit to recreate $|\psi\rangle$ when $(M_1, M_2) = (1,0)$ or $(M_1, M_2) = (1,1)$, respectively.

Quantum teleportation has been experimentally demonstrated,[14] confirming that quantum correlation (entanglement) is nonlocal; the effect is independent of the distance since Alice does not need to know where Bob is. It is important to emphasize, however, that this successful teleportation does not indicate faster-than-light communication, since classical communication is involved. Alice has to deliver the classical bits (M_1 and M_2) to Bob by some classical means, so relativity is not violated.

[14] See, e.g., D. Bouwmeester *et al.*, *Nature* **390**, 575 (1997); D. Boschi *et al.*, *Physical Review Letters* **80**, 1121 (1998); A. Furusawa *et al.*, *Science* **282**, 706 (1998); I. Marcikic *et al.*, *Nature* **421**, 509 (2003).

4.6 Quantum Algorithms

Now that we are familiar with qubits, quantum gates, and quantum circuits, let us use these tools and explore some representative examples of quantum algorithms, in order to understand how quantum computation can be faster and more powerful than its classical counterparts in solving certain problems.

4.6.1 Quantum Parallelism

We consider a quantum circuit for evaluating a function f, depicted in Figure 4.11. Here, f is a Boolean function, so both the input and

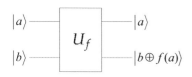

Figure 4.11 Quantum circuit for evaluating the values of an arbitrary Boolean function $f: \{0,1\} \to \{0,1\}$.

output are binary numbers, taking values of either 0 or 1. The box in the middle, denoted as U_f, represents a two-qubit quantum gate; the time evolution of the first qubit is shown at the top, and that of the second qubit is shown at the bottom, as usual. As can be seen in the figure, the first qubit, $|a\rangle$, does not change, just like the control qubit in a CNOT gate, whereas the second qubit changes through addition modulo 2 by the function f evaluated at a, i.e., the value of the first qubit.

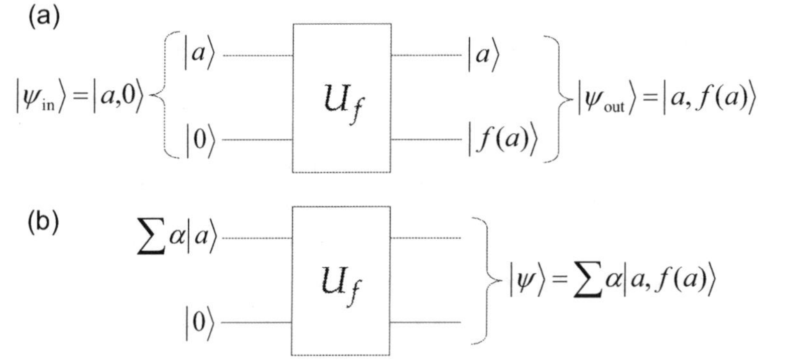

(a)

$$|\psi_{\text{in}}\rangle = |a,0\rangle \left\{ \begin{array}{c} |a\rangle \\ |0\rangle \end{array} \right. \boxed{U_f} \left. \begin{array}{c} |a\rangle \\ |f(a)\rangle \end{array} \right\} |\psi_{\text{out}}\rangle = |a,f(a)\rangle$$

(b)

$$\sum \alpha |a\rangle \quad \boxed{U_f} \quad \left. \right\} |\psi\rangle = \sum \alpha |a,f(a)\rangle$$
$$|0\rangle$$

Figure 4.12 The principle of parallel evaluations of function f. (a) A special case of the function-evaluation quantum circuit shown in Figure 4.11 where the second input qubit $|b\rangle$ is $|0\rangle$. (b) Generalization of (a) so that a superposition state is the first input state. The resulting output state is also a superposition state with different terms containing the function f evaluated at multiple values simultaneously, using the quantum parallelism of superposition states.

If the second input qubit $|b\rangle$ happens to be zero, then it becomes simply $|f(a)\rangle$ after going through this circuit; see Figure 4.12(a). In other words, if the input two-qubit state is $|a,0\rangle$ then the output two-qubit state is $|a,f(a)\rangle$. Since f is a binary function, its value is either 0 or 1. By simple generalization, we can input a single-qubit superposition state for the first qubit, $\sum \alpha |a\rangle$, with expansion coefficients α. Then we find that the output two-qubit state is also a superposition state, $\sum \alpha |a,f(a)\rangle$.

To be more specific, let us use a 50%–50% superposition of $|0\rangle$ and $|1\rangle$, $\frac{|0\rangle+|1\rangle}{\sqrt{2}}$, as the input for the first qubit. Then the output two-qubit state is a 50%–50% superposition of $|0,f(0)\rangle$ and $|1,f(1)\rangle$. Note that $f(0)$ and $f(1)$ appear *simultaneously* through just one operation. The implication is that a single quantum circuit can be used to evaluate the function $f(x)$ for multiple values of x simultaneously, owing to the ability of a quantum computer to be in superposition states. This property is called quantum parallelism.

We can generalize the process of parallel evaluations to n-qubit systems. The first step is to create an equal superposition of all n basis states using the Walsh–Hadamard transform, $H^{\otimes n}$, which is shown in Figure 4.13. The output state of this procedure is

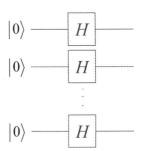

Figure 4.13 The Walsh–Hadamard transform, $H^{\otimes n}$, of an initial state $|0\rangle^{\otimes n}$ to create an equal superposition of all n basis states $(1/\sqrt{2^n}) \sum_x |x\rangle$.

$$|\psi_{\text{out}}\rangle = \frac{(|0\rangle+|1\rangle)(|0\rangle+|1\rangle)\cdots(|0\rangle+|1\rangle)}{\sqrt{2^n}} = \frac{1}{\sqrt{2^n}} \sum_x |x\rangle, \quad (4.42)$$

where x in the summation runs through all possible n-bit values. For $n = 1$, we recover the usual Hadamard gate, which is a single-qubit gate and produces $\frac{|0\rangle + |1\rangle}{\sqrt{2}}$. For $n = 2$, we see that the Walsh–Hadamard transform is a two-qubit gate and produces an equal superposition of the four basis states, $(|0\rangle + |1\rangle)(|0\rangle + |1\rangle)/\sqrt{2^2} = (|00\rangle + |01\rangle + |10\rangle + |11\rangle)/2$.

Here is a step-by-step procedure for demonstrating parallel evaluations of a function f at n multiple values, following the quantum circuit shown in Figure 4.14. The first step is to prepare an initial $(n + 1)$-qubit state, $|\psi_0\rangle = |0\rangle^{\otimes n}|0\rangle$. The next step is to apply the Walsh–Hadamard transform of nth order, $H^{\otimes n}$, to the first n qubits, which produces a superposition state

$$|\psi_1\rangle = \frac{1}{\sqrt{2^n}} \sum_x |x\rangle|0\rangle. \tag{4.43}$$

In the final step, we apply U_f, as shown in Figure 4.11, which produces a superposition of states with f evaluated at n different x values simultaneously:

$$|\psi_2\rangle = \frac{1}{\sqrt{2^n}} \sum_x |x, f(x)\rangle. \tag{4.44}$$

So, this procedure shows that a single operation can evaluate all possible values of f simultaneously, demonstrating the core idea of quantum parallelism.

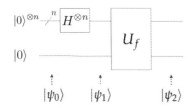

Figure 4.14 Quantum circuit for demonstrating parallel evaluations of function f at n multiple values. $H^{\otimes n}$ is the Walsh–Hadamard gate defined in Figure 4.13, and the function evaluator U_f is defined in Figure 4.11.

4.6.2 Deutsch's Algorithm

The results we have just obtained in the last subsection are fascinating, but we have to be cautious as to what we can measure because quantum measurement is subtle and nontrivial.[15] Even though we have a superposition state that contains information on f evaluated at different x values simultaneously, this does not automatically mean that we can measure all those terms simultaneously. Quantum measurement will project the state onto only one of the eigenstates,[16] thus allowing us to determine only one term at a time. Furthermore, this comes with a finite probability as to which state will be measured, so a certain randomness is introduced. In other words, quantum information is hidden in the superposition. For quantum parallelism to be useful, we have to find a way to extract information about more than one value of $f(x)$ from the superposition state.

In 1985, David Deutsch proposed an algorithm which realizes exactly this.[17] The algorithm is implemented by the quantum circuit depicted in Figure 4.15. This is a two-qubit quantum circuit, and we

[15] See Section 3.5.

[16] The phenomenon of wavefunction collapse.

[17] D. Deutsch, *Proceedings of the Royal Society of London A* **400**, 97 (1985).

start with $|0\rangle$ and $|1\rangle$ for the first and second qubits, respectively. Each of these initial single-qubit states goes through a Hadamard gate, and they enter the function evaluator U_f.[18] After that, the first qubit goes through a Hadamard gate one more time. We denote the state vector at different times as $|\psi_0\rangle$, $|\psi_1\rangle$, $|\psi_2\rangle$, and $|\psi_3\rangle$ as it evolves in time while going through this circuit.

The initial input state $|\psi_0\rangle = |01\rangle$ becomes

$$|\psi_1\rangle = \left(\frac{|0\rangle + |1\rangle}{\sqrt{2}}\right)\left(\frac{|0\rangle - |1\rangle}{\sqrt{2}}\right) \tag{4.45}$$

after each single qubit goes through a Hadamard gate. Then, after going through U_f, the state vector will have one of the two forms, depending on whether $f(0)$ and $f(1)$ are equal or not equal:

$$|\psi_2\rangle = \begin{cases} \pm\left(\frac{|0\rangle + |1\rangle}{\sqrt{2}}\right)\left(\frac{|0\rangle - |1\rangle}{\sqrt{2}}\right), & f(0) = f(1), \\[2ex] \pm\left(\frac{|0\rangle - |1\rangle}{\sqrt{2}}\right)\left(\frac{|0\rangle - |1\rangle}{\sqrt{2}}\right), & f(0) \neq f(1). \end{cases} \tag{4.46}$$

In the final step, the first qubit is transformed back into $|0\rangle$ and $|1\rangle$, respectively, by the Hadamard gate. Therefore, the final state is

$$|\psi_3\rangle = \begin{cases} \pm|0\rangle\left(\frac{|0\rangle - |1\rangle}{\sqrt{2}}\right), & f(0) = f(1), \\[2ex] \pm|1\rangle\left(\frac{|0\rangle - |1\rangle}{\sqrt{2}}\right), & f(0) \neq f(1), \end{cases}$$
$$= \pm|f(0) \oplus f(1)\rangle\left(\frac{|0\rangle - |1\rangle}{\sqrt{2}}\right). \tag{4.47}$$

In the final step, we used the fact that if $f(0) = f(1)$ then $f(0) \oplus f(1) = 0$, and if $f(0) \neq f(1)$, then one of $f(0)$ and $f(1)$ is 0 and the other is 1, so that $f(0) \oplus f(1) = 1$. Therefore, the information on whether $f(0) = f(1)$ or $f(0) \neq f(1)$ is contained in the first qubit, and hence, if we measure the first qubit, we immediately know the value of $f(0) \oplus f(1)$. This is already an impressive achievement since we have been able to determine the sum of $f(0)$ and $f(1)$ through one operation, whereas a classical computer would have required two operations to determine the sum of the two since that would require a knowledge of the individual values.

4.6.3 Quantum Advantages

What we have just demonstrated in the preceding subsection using Deutsch's algorithm is a simple but definite advantage of quantum computation over classical computation, sometimes referred to as a

[18] See Figure 4.11 to see how U_f works.

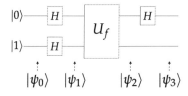

Figure 4.15 Quantum circuit for implementing Deutsch's algorithm. © Deyin Kong (Rice University).

[19] For more details about quantum algorithms and their advantages, see, e.g., M. A. Nielsen and I. L. Chuang, *Quantum Computation and Quantum Information* (Cambridge University Press, 2000), and S. Aaronson, *Quantum Computing since Democritus* (Cambridge University Press, 2013).

quantum advantage. Below, let us look at a few other examples of quantum algorithms[19] that are expected to lead to quantum advantages or, to be more specific, a massive acceleration or speedup in computation if they could be successfully implemented in real physical devices. Rather than detailing how these algorithms work, we will summarize their main advantages, by highlighting how the difficulty of computation grows as the number of required operations or steps increases, as compared with classical computation applied to the same problems.

The Deutsch–Jozsa algorithm. In 1992, David Deutsch and Richard Jozsa extended Deutsch's algorithm to the n-qubit case.[20] Its quantum circuit is depicted in Figure 4.16. This algorithm can show quantum acceleration in a specific problem, known as Deutsch's problem, as outlined below. Alice selects a number x from 0 to $2^n - 1$ and sends it to Bob, who then calculates $f(x)$ and replies with the result, which is 0 or 1. The function $f(x)$ is assumed to be special in the sense that it is either constant[21] or balanced,[22] and Alice would like to determine whether f is constant or balanced with minimal communication. Classically, in the worst case Alice needs to ask Bob $2^{n-1} + 1$ times. If we use Deutsch's and Jozsa's quantum algorithm, by starting with the initial $(n+1)$-qubit state $|\psi_0\rangle$ and going through the circuit step by step, we obtain

[20] D. Deutsch and R. Jozsa, *Proceedings of the Royal Society of London A* **439**, 553 (1992).

[21] Which means that $f(x)$ is always 0 or always 1 no matter what x is.

[22] Which means that $f(x)$ is 1 for half the input and 0 for the other half.

$$|\psi_3\rangle = \sum_z \sum_x \frac{(-1)^{xz+f(x)}|z\rangle}{2^n} \left(\frac{|0\rangle - |1\rangle}{\sqrt{2}} \right). \qquad (4.48)$$

Note that in the summation over z, the term that corresponds to $|0\rangle^{\otimes n}$ occurs when $z = 0$. Therefore, the amplitude for that term is given by $\sum_x (-1)^{f(x)}/2^n$, and hence the probability of measuring $|0\rangle^{\otimes n}$ is $|\sum_x (-1)^{f(x)}/2^n|^2$, which is 1 if $f(x)$ is constant and 0 if $f(x)$ is balanced. In other words, the final measurement by Alice on the n-qubit state will be all zeros if $f(x)$ is constant and will lead to some other states if $f(x)$ is balanced. It should be emphasized that Deutsch's problem can be solved with one evaluation of f by using the Deutsch–Jozsa algorithm, in contrast with the classical requirement of $2^{n-1} + 1$ evaluations.

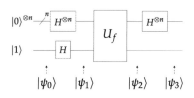

Figure 4.16 Quantum circuit for implementing the Deutsch–Jozsa algorithm, which is an extension of Deutsch's algorithm (see Figure 4.15) to the case of n qubits. © Deyin Kong (Rice University).

Grover's search algorithm. Another type of quantum speedup can be demonstrated by Grover's algorithm,[23] also known as the quantum search algorithm. Suppose that we want to find one particular item with a unique property by searching a database of size N. The database has no structure, and the items in it are not sorted in any

[23] L. K. Grover, in: *Proceedings of the 28th Annual Association for Computing Machinery (ACM) Symposium on Theory of Computing* (ACM Press, New York, 1996), pp. 212–219; DOI: 10.1145/237814.237866; see also *Physical Review Letters* **97**, 325 (1997).

way. Classically, this problem requires approximately N queries, because, in the worst case scenario, there is a possibility that we cannot find the item after $N - 1$ queries but then we still have to check the Nth item. On average, finding the item would still take $N/2$ queries, that is, the number of queries is linear in N. However, Grover's algorithm allows one to solve this problem using approximately $\sqrt{N}$ operations. This speedup compared to classical computers ($\sim N$ versus $\sim \sqrt{N}$) is therefore called a quadratic speedup. This can be a considerable speedup when N is large, and Grover's algorithm can have diverse applications, even though it is a modest speedup compared to the *exponential* speedup that we see next in Shor's algorithm.

Shor's factorization algorithm. The most commonly used method in cryptography is the RSA (Rivest–Shamir–Adleman) scheme, which relies on the practical difficulty of factorization. It is well known that factorizing a large integer is computationally intractable, and no classical algorithms currently exist that can factor integers in polynomial time. In 1994, Peter Shor reported a quantum algorithm that enables efficient integer factorization on an ideal quantum computer.[24] This work has thus raised the possibility that RSA would be defeated if a large quantum computer could be constructed. Further, much excitement was generated owing to the promise given by this algorithm that one day a real-world impact will be made by quantum computers (which was not the case with the Deutsch–Jozsa algorithm). A significant amount of effort followed Shor's work in designing and developing novel quantum algorithms, while at the same time research on new cryptosystems based on quantum cryptography began. Hence, today's quantum evolution has much to owe to Shor's seminal work. The efficiency of Shor's algorithm can be attributed to the efficiency of the quantum Fourier transform.[25] In the case of an optimized classical fast Fourier transform algorithm, the number of logic gates required for transforming an n-bit number is $\sim n2^n$, while, in the corresponding discrete quantum Fourier transform, the required number of quantum gates is only $n(n+1)/2 \sim n^2$, so there is an exponential speedup.

Quantum simulation. Simulating quantum systems with a computer is a notoriously difficult task especially when the system size is large. Just to store full information on the state of the system requires too much computer memory. For example, we know that there are two amplitudes, α and β (where $|\alpha|^2 + |\beta|^2 = 1$), to store for specifying the state of a single two-level quantum particle. When we have two such particles, the number of amplitudes to store becomes 4, and when

[24] P. W. Shor, in: *Proceedings of the 35th Annual Symposium on Foundations of Computer Science* (IEEE Computer Society Press, Los Alamitos, California, 1994), p. 24.

[25] Both Shor's algorithm and the Deutsch–Jozsa algorithm are based on quantum versions of the Fourier transform.

[26] And when we have a system of 100 particles, we have $2^{100} \approx 10^{30}$ amplitudes to store. Further, we have to keep track of them as a function of time if we are simulating the dynamics of the system.

[27] "Let the computer itself be built of quantum mechanical elements which obey quantum mechanical laws," said Richard Feynman [*International Journal of Theoretical Physics* **21**, 467 (1982)].

[28] I. M. Georgescu, S. Ashhab, and F. Nori, "Quantum simulation," *Reviews of Modern Physics* **86**, 153 (2014); E. Altman *et al.*, "Quantum simulators: Architectures and opportunities," *PRX Quantum* **2**, 017003 (2021).

[29] A *universal* quantum computer should have an ensemble of well-defined qubits that can be initialized, controlled, and measured, and it should be able to tackle a variety of different problems, unlike special-purpose quantum computers. See, e.g., S. Lloyd, *Science* **273**, 1073 (1996).

[30] For more specific requirements, see, e.g., D. P. DiVincenzo, *Quantum Information and Computation* **1**, 1 (2001), and S. Aaronson, *Quantum Computing since Democritus* (Cambridge University Press, 2013).

[31] For a recent review, see, e.g., N. P. de Leon *et al.*, *Science* **372**, eabb2823 (2021).

we have 10 particles, $2^{10} \approx 10^3$ amplitudes must be stored.[26] This *exponential explosion* problem can be circumvented by using quantum simulators instead of classical simulators; we would then need only N qubits (as opposed to 2^N classical bits) to model an N-particle system. As quantum systems, quantum simulators can provide much more insight into quantum phenomena than classical simulators.[27] Hence, the basic idea of quantum simulation is to use engineered quantum matter with well-understood and controllable microphysics and good diagnostics (e.g., ultracold atoms or semiconductor nanodevices) to provide insight into complex quantum systems and processes. Quantum simulators can be viewed as special-purpose quantum computers designed to solve or discover aspects of macroscopic systems that possess intractably large degrees of freedom.[28] Quantum simulation promises to advance our understanding of important open questions in materials and chemistry, which will lead to the creation of advanced materials with designer properties as well as having applications in the study of many problems in condensed matter physics, high-energy physics, atomic physics, quantum chemistry, and cosmology. It is especially promising for unambiguously providing quantum advantages without waiting for fully fledged universal quantum computers[29] to be built (which may not happen for decades). Just as analog classical computers served as a bridge to digital classical computers, it is likely that quantum simulators will provide the first transformative results of the second quantum revolution by unravelling the secrets of complex quantum systems.

4.7   *Quantum Hardware*

There is a variety of two-level systems that can be used as qubits, including the polarization states of photons, the spin states of electrons and neutrons, and the orbital states of real and artificial atoms. In a typical quantum information processing protocol, a qubit is prepared in a certain superposition state ("initialization"), evolves under an appropriate unitary transformation ("operation"), and is then measured to produce the result ("measurement"). The goal[30] is to find a system that allows one to perform as many operations as possible within the coherence time. Scalability is another important question if one wants a quantum computer to have real-world impact. Therefore, various atomic and molecular systems as well as solid-state materials and devices are currently being explored, as summarized below.[31]

Bulk NMR quantum computers. Utilization of the long-lived quantum coherence of nuclear spins for quantum computing was theoretically proposed by Seth Lloyd[32] and David DiVincenzo,[33] and nuclear magnetic resonance (NMR) quantum computing thus became one of the oldest approaches for constructing a quantum computer. This approach utilizes the spin states of nuclei within molecules as qubits, and the quantum states are probed through the use of standard techniques available in NMR spectroscopy. This type of quantum hardware consists of an ensemble of molecules, rather than a single pure state, which is why this technique is called *bulk* NMR quantum computation. Experimental implementations were achieved by Cory and coworkers and Gershenfeld and Chuang in 1997.[34] In 2001, a group at IBM reported the successful implementation of Shor's algorithm (Section 4.6) in a seven-qubit NMR quantum computer.[35]

Trapped ions. In trapped-ion platforms for quantum information processing, ions can be confined and suspended in free space using AC electromagnetic fields, typically through the use of a technique called the Paul trap.[36] A qubit exists in the electronic states of each ion, and quantum information can be transferred through interaction with the collective quantized motion of the ions. Laser beams with appropriately designed wavelengths, intensities, and durations are used for controlling the quantum states and dynamics by performing unitary operations. The first implementation scheme for a controlled-NOT quantum gate was proposed by Ignacio Cirac and Peter Zoller,[37] and a number of groups around the world have implemented the idea. This is one of the most promising architectures for a scalable, universal quantum computer today, and commercial machines are already available. The highest operation fidelities of any quantum system have been achieved in trapped-ion systems, and successful quantum simulation has been performed of spin models with variable-range interactions using more than 50 spins.

Cavity QED systems. Cavity quantum electrodynamics (QED) refers to the study of light–matter interaction in a cavity setting, where the quantum nature of light is significant. Specifically, the quantum states and dynamics of a single two-level atom resonantly interacting with single-mode cavity photons, depicted in Figure 4.17(b), can be described in cavity QED through the Jaynes–Cummings model.[38] The system can undergo many cycles of Rabi oscillations (Section 6.1) in the strong coupling regime where the light–matter coupling rate exceeds the decay rates of light and matter. This system can be used as a single-photon source (starting with an excited atom), or as an interface between an atom and an optical network. By adjusting the interaction duration, one can also create entanglement between the

[32] S. Lloyd, *Science* **261**, 1569 (1993).

[33] D. DiVincenzo, *Physical Review A* **51**, 1015 (1995).

[34] D. G. Cory *et al.*, *Proceedings of the National Academy of Sciences* **94**, 1634 (1997); N. A. Gershenfeld and I. L. Chuang, *Science* **275**, 350 (1997).

[35] L. M. Vandersypen *et al.*, *Nature* **414**, 883 (2001).

(a)

(b)

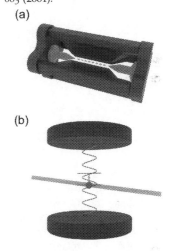

Figure 4.17 (a) Trapped ion quantum computer using a Paul trap. See, e.g., G. Pagano *et al.*, *Proceedings of the National Academy of Sciences of the USA* **117**, 25396 (2020). (b) Cavity QED systems are ideally suited for studying the interaction of quantized light and quantized matter. See, e.g., P. Domokos *et al.*, *Physical Review A* **52**, 3554 (1995). © Deyin Kong (Rice University).

[36] Also known as the quadrupole ion trap, it uses radio-frequency electric fields to trap charged particles. Wolfgang Paul shared the Nobel Prize in Physics in 1989 for inventing this technique. See Figure 4.17(a).

[37] I. J. Cirac and P. Zoller, *Physical Review Letters* **74**, 4091 (1995).

[38] E. T. Jaynes and F. W. Cummings, *Proceedings of IEEE* **51**, 89 (1963).

atom and cavity field. In addition to cavity QED systems based on atoms or molecules, semiconductor-based cavity QED systems have emerged[39] which have exhibited *ultrastrong* coupling owing to the enormous dipole moments characteristic of transitions in solids.

Superconducting qubits and circuit QED systems. Superconducting quantum computing is a highly promising solid-state implementation of quantum computation based on superconducting electronic circuits, or circuit QED.[40] There are currently intensive world-wide research efforts on superconducting qubits by a large number of groups both in academia and industry, including major high-tech companies. In analogy to cavity QED, circuit QED is concerned with studying the interaction between quantized light and quantized matter, which is describable by the Jaynes–Cummings model, especially for quantum information processing purposes. In contrast with cavity QED, however, the photon (with energies usually in the microwave range) is stored in a 1D on-chip resonator and the quantum object is an artificial atom, namely, a superconducting qubit; see, e.g., Figure 4.18. There are three general types of superconducting qubits – phase, charge, and flux qubits – depending on which degrees of freedom of the superconducting states are utilized; their hybridizations also exist. Any superconducting quantum circuit contains a Josephson junction,[41] which introduces a nonlinear inductance, making an anharmonic oscillator, as is needed to create a qubit.[42] In 2019, a group at Google led by John Martinis[43] built quantum processors based on 54 superconducting qubits and performed computations in a Hilbert space of dimension $2^{53} \approx 10^{16}$ (one of the qubits did not function well, which is why this number is 2^{53}, instead of 2^{54}), which is beyond the reach of the fastest classical supercomputers available today.

[39] N. Marquez Peraca *et al.*, in: *Semiconductor Quantum Science and Technology*, Semiconductors and Semimetals **105** (Elsevier, 2020), pp. 89–151.

[40] See, e.g., A. Blais *et al.*, "Cavity quantum electrodynamics for superconducting electrical circuits: An architecture for quantum computing," *Physical Review A* **69**, 062320 (2004); J. Q. You and F. Nori, "Superconducting circuits and quantum information," *Physics Today* **58**, 42 (2007).

[41] See Exercise 5.9.

[42] A harmonic oscillator cannot be used as a qubit, or a two-level system, because it posses an infinite ladder of equally spaced states; see Figure 3.1.

[43] F. Arute *et al.*, *Nature* **574**, (2019).

Figure 4.18 One type of superconducting qubit called a "transmon qubit" coupled to a 1D transmission-line resonator. In the middle, two superconducting islands are connected by a Josephson junction. See, e.g., A. Blais, S. M. Girvin, and W. D. Oliver, *Nature Physics* **16**, 247 (2020) for more details about superconducting circuit QED. © Deyin Kong (Rice University).

Quantum dots and quantum defects in semiconductors. There are many different types of solid-state approaches to the realization of scalable quantum information processing. Solid-state systems, as opposed to atomic and molecular systems, are preferred because solids can offer a much greater degree of control over design and fabrication, which is needed for constructing large-scale devices. In addition to the approaches based on superconducting circuits that we have just seen, there are many semiconductor-based approaches. Semiconductor quantum dots[44] exhibit very sharp, atom-like spectral features in emission, scattering, and absorption,[45] implying long-lived quantum coherence. Time-domain studies of quantum dots of various types have demonstrated coherent control of quantum states, including, most importantly, Rabi oscillations,[46] which in turn demonstrates that populations of electrons in quantum dots can be coherently manipulated just as in real atoms driven by coherent external fields (see Section 6.1). CNOT gate operations have also been demonstrated in quantum dots.[47] More recently, point defects in semiconductors, especially deep centers in wide-gap semiconductors (or insulators), have emerged as robust atom-like quantum centers in solids.[48] Qubits based on these so-called quantum defects can be initialized, manipulated, and read out, often at room temperature. They have also been successfully demonstrated as single-photon emitters. A prime example of this class of quantum defects is the nitrogen-vacancy center in diamond, but there are many other types of defects reported for diverse materials with a range of characteristic wavelengths, including SiC, carbon nanotubes, boron nitride, and transition metal dichalcogenides. There is currently active research that involves sorting through candidate defect systems to achieve superior performance or greater functionality.

[44] See Section 7.3.

[45] See, e.g., M. Grundmann *et al.*, *Physical Review Letters* **74**, 4043 (1995); D. Gammon *et al.*, *Physical Review Letters* **76**, 3005 (1996); and H. D. Robinson and B. B. Goldberg, *Physical Review B* **61**, R5086 (2000).

[46] See, e.g., T. H. Stievater *et al.*, *Physical Review Letters* **87**, 133603 (2001); H. Kamada *et al.*, *Physical Review Letters* **87**, 246401 (2001); H. Htoon *et al.*, *Physical Review Letters* **88**, 087401 (2002); A. Zrenner *et al.*, *Nature* **418**, 612 (2002); Z. Shi *et al.*, *Physical Review B* **88**, 075416 (2013).

[47] X. Li *et al.*, *Science* **301**, 809 (2003).

[48] See, e.g., J. R. Weber, *Proceedings of the National Academy of Sciences of the USA* **107**, 8513 (2010); M. W. Doherty *et al.*, *Physics Reports* **528**, 1 (2013); L. Gordon *et al.*, *MRS Bulletin* **38**, 802 (2013); J. F. Barry *et al.*, *Reviews of Modern Physics* **92**, 015004 (2020); X. He *et al.*, *Nature Photonics* **11**, 577 (2017); R. Bourrellier *et al.*, *Nano Letters* **16**, 4317 (2016); and C. Palacios-Berraquero *et al.*, *Nature Communications* **7**, 12978 (2016).

4.8   *Chapter Summary*

This chapter has provided an introduction to the subject of quantum information science, which is one of the most important emerging applications of quantum mechanics. Quantum information is encoded as qubits in a 2D Hilbert space with basis states $|0\rangle$ and $|1\rangle$. Any superposition of them, $\alpha|0\rangle + \beta|1\rangle$, can be represented as a unit vector (known as the Bloch vector) in the Hilbert space whose tip lies on the surface of the Bloch sphere (Figure 4.1). A quantum gate operation on such a single qubit is a unitary operation and can be visualized as a rotation of the Bloch vector. A special type of correlations between quantum particles – quantum entanglement – can occur when multiple qubits exist. The most representative two-qubit entangled states are the Bell states [Equations (4.15)–(4.18)], which play crucial roles in quantum technologies. In an entangled state, joint properties of different qubits are well defined whereas individual qubits do not carry any well-defined information. Multiple-qubit gates can be represented by a combination of single-qubit gates and the CNOT gate. The time evolution of a qubit system can be most conveniently described by a quantum circuit. Some major components found in quantum circuits are shown in Figure 4.6, and examples of quantum circuits are provided in Examples 4.1 and 4.2 as well as in Section 4.5. Some examples of quantum algorithms and quantum hardware are described in Sections 4.6 and 4.7, respectively. Overall, becoming familiar with the definitions and methods described in this chapter is fundamentally important for understanding the more complicated algorithms that are in practice used in quantum computation and communications.

4.9 Exercises

Exercise 4.1 (Different bases)

Let $\{|0\rangle, |1\rangle\}$ be an orthonormal basis set in a 2D Hilbert space. The NOT operation (unitary operator) is defined as

$$|0\rangle \rightarrow |1\rangle, \quad |1\rangle \rightarrow |0\rangle. \tag{4.49}$$

(a) Find the unitary operator $\hat{U}_{\text{NOT}}$ that implements the NOT operation with respect to the basis $\{|0\rangle, |1\rangle\}$.

(b) Let

$$|0\rangle = \begin{bmatrix} 1 \\ 0 \end{bmatrix}, \qquad |1\rangle = \begin{bmatrix} 0 \\ 1 \end{bmatrix}. \tag{4.50}$$

Find the matrix representation of $\hat{U}_{\text{NOT}}$ for this basis.

(c) Let

$$|0\rangle = \frac{1}{\sqrt{2}} \begin{bmatrix} 1 \\ 1 \end{bmatrix}, \qquad |1\rangle = \frac{1}{\sqrt{2}} \begin{bmatrix} 1 \\ -1 \end{bmatrix}. \tag{4.51}$$

Find the matrix representation of $\hat{U}_{\text{NOT}}$ for this basis.

Exercise 4.2 (Tensor or Kronecker product)

Use the basis given by Equations (4.2) and (4.3), when needed, in answering the following questions:

(a) Calculate $|1\rangle \otimes |0\rangle$ and $|1\rangle \otimes |1\rangle$.

(b) For $|\psi\rangle = \frac{1}{\sqrt{2}}(|0\rangle + |1\rangle)$, calculate $|\psi\rangle^{\otimes 2}$ and $|\psi\rangle^{\otimes 3}$.

(c) For the Pauli matrices σ_x and σ_z, calculate $\sigma_x \otimes \sigma_z$ and $\sigma_z \otimes \sigma_x$.

(d) For the Hadamard matrix $\mathbf{H}$,[49] calculate $\mathbf{H}^{\otimes 2}$.   [49] See Equation (4.31).

Exercise 4.3 (Entanglement)

(a) Show that

$$|\psi\rangle = \frac{1}{\sqrt{2}}(|01\rangle - |10\rangle) \tag{4.52}$$

is entangled by demonstrating that it *cannot* be written as a product state such as $(c_0|0\rangle + c_1|1\rangle) \otimes (d_0|0\rangle + d_1|1\rangle)$.

(b) Show that

$$|\psi\rangle = x_1|00\rangle + x_2|01\rangle + x_3|10\rangle + x_4|11\rangle \tag{4.53}$$

is disentangled (i.e., separable) if and only if $x_1 x_4 = x_2 x_3$.

(c) Consider the following unitary operator in a 2D Hilbert space:

$$\mathbf{U}(\theta,\phi) = \begin{bmatrix} \cos(\theta/2) & e^{-i\phi}\sin(\theta/2) \\ -e^{-i\phi}\sin(\theta/2) & \cos(\theta/2) \end{bmatrix}. \tag{4.54}$$

Which of the following states are entangled?

(i) $(\mathbf{U}(\theta_1,\phi_1) \otimes \mathbf{U}(\theta_2,\phi_2))\,[1,0,0,0]^T$,

(ii) $(\mathbf{U}(\theta_1,\phi_1) \otimes \mathbf{U}(\theta_2,\phi_2))\,[0,0,0,1]^T$,

(iii) $(\mathbf{U}(\theta_1,\phi_1) \otimes \mathbf{U}(\theta_2,\phi_2))\,\frac{1}{\sqrt{2}}[1,0,0,1]^T$,

where ^{T} denotes the transpose.

Exercise 4.4 (Noon state)

The following entangled state is called the "noon" state (or, more appropriately, the "$n00n$" state):

$$|\psi_{\text{noon}}(\theta)\rangle = \frac{|n,0\rangle + e^{in\theta}|0,n\rangle}{\sqrt{2}}. \tag{4.55}$$

Here, $|n,0\rangle = |n\,0\rangle = |n\rangle|0\rangle$ is a product state of two number states[50] with n photons in the first state and 0 photons in the second state, and vice versa for $|0,n\rangle = |0\,n\rangle = |0\rangle|n\rangle$. Answer the following questions for the observable A represented by

$$\hat{A} := |n,0\rangle\langle 0,n| + |0,n\rangle\langle n,0|. \tag{4.56}$$

(a) Show that $\langle A\rangle_\theta = \langle \psi_{\text{noon}}(\theta)|\hat{A}|\psi_{\text{noon}}(\theta)\rangle = \cos(n\theta)$, where θ is the phase.

(b) Calculate $\langle A^2\rangle_\theta$.

(c) Evaluate the variance $\Delta A = \sqrt{\langle A^2\rangle_\theta - \langle A\rangle_\theta^2}$.

(d) Show that $\Delta\theta = 1/n$ and thus the larger is n, the more precise is θ. Hint: $\Delta\theta = \Delta A\,|d\langle A\rangle_\theta/d\theta|^{-1}$.

Exercise 4.5 (Single-qubit gates)

(a) The Z gate transforms $|0\rangle$ and $|1\rangle$ as

$$|0\rangle \rightarrow |0\rangle, \tag{4.57}$$
$$|1\rangle \rightarrow -|1\rangle. \tag{4.58}$$

Obtain the matrix representation of the gate $\mathbf{Z}$ using the basis given by Equations (4.2) and (4.3). Show that $\mathbf{Z}^2 = \mathbf{I}_{2D}$.

(b) Show that the Hadamard gate can be viewed as a rotation about the y axis by $90°$ followed by a reflection through the x–y plane. *Hint*: See Equation (4.27).

(c) The phase shifter transforms $|0\rangle$ and $|1\rangle$ as

$$|0\rangle \to e^{i\phi}|0\rangle, \tag{4.59}$$

$$|1\rangle \to |1\rangle. \tag{4.60}$$

Obtain a matrix representation of this gate, Φ, using the basis given by Equations (4.2) and (4.3). Show that $\Phi^\dagger\Phi = I_{2D}$.

(d) Obtain the matrix representation of the gate, $H\Phi H$, using the basis given by Equations (4.2) and (4.3). Here H is the Hadamard matrix.[51] Explain how this gate transforms $|0\rangle$ and $|1\rangle$. When $|\phi_{in}\rangle = |0\rangle$, show that $\langle\phi_{out}|\phi_{out}\rangle = 1$.

[51] See Equation (4.31).

Exercise 4.6 (Time evolution)

Consider the linear operator

$$\hat{H} = i\hbar\omega(|0\rangle\langle 1| - |1\rangle\langle 0|) \tag{4.61}$$

operating in the 2D Hilbert space formed by the basis $\{|0\rangle, |1\rangle\}$.

(a) Show that $\hat{H}$ is hermitian, i.e., $\hat{H}^\dagger = \hat{H}$.

(b) Find the eigenvalues and corresponding normalized eigenvectors of $\hat{H}$.

(c) Calculate the time evolution operator $\hat{U}(t) = \exp(-i\hat{H}t/\hbar)$.[52] Find the value of t such that $\hat{U}(t) = |0\rangle\langle 1| - |1\rangle\langle 0|$.

[52] See Equation (2.81).

(d) Calculate $\hat{U}(t = \pi/4\omega)$ and $(\hat{U}(t = \pi/4\omega))^2$.

Exercise 4.7 (Two-qubit gates)

(a) Derive Equation (4.32).

(b) Prove that the controlled-NOT gate can be written as $\hat{U}_{CNOT} = |0\rangle\langle 0| \otimes \hat{I} + |1\rangle\langle 1| \otimes \hat{U}_{NOT}$, where $\hat{I}$ is the 2D identity operator.

(c) Show that $(\hat{U}_H \otimes \hat{U}_H)\hat{U}_{CNOT}(\hat{U}_H \otimes \hat{U}_H)|j,k\rangle = |j \oplus k, k\rangle$, where $|j,k\rangle = |j\rangle \otimes |k\rangle$ with $j,k \in \{0,1\}$. Here, $\hat{U}_H$ is the Hadamard gate, and $\hat{U}_{CNOT}$ is the CNOT gate.

(d) Let $\{|0\rangle, |1\rangle\}$ be an orthonormal basis in a 2D Hilbert space and

$$\hat{U}_{PS}(\theta) := |00\rangle\langle 00| + |01\rangle\langle 01| + |10\rangle\langle 10| + e^{i\theta}|11\rangle\langle 11|. \tag{4.62}$$

(i) Show that

$$(\hat{I}_2 \otimes \hat{U}_{\mathrm{H}})\hat{U}_{\mathrm{PS}}(\pi)(\hat{I} \otimes \hat{U}_{\mathrm{H}})|AB\rangle = |A, A \oplus B\rangle. \qquad (4.63)$$

(ii) Show that

$$(\hat{I}_2 \otimes \hat{U}_{\mathrm{H}})\hat{U}_{\mathrm{CNOT}}(\hat{I} \otimes \hat{U}_{\mathrm{H}})|AB\rangle = (-1)^{AB}|AB\rangle. \qquad (4.64)$$

Exercise 4.8 (The Toffoli gate)

Figure 4.19 The quantum circuit that defines the quantum Toffoli gate.

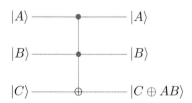

The quantum circuit for the Toffoli gate is shown in Figure 4.19. It is a three-qubit gate with input states $|A\rangle$, $|B\rangle$, and $|C\rangle$.

(a) Construct the truth table for the Toffoli gate.

(b) Express the unitary operator representing the Toffoli gate, $\hat{U}_{\mathrm{Toffoli}}$, in terms of the basis kets of an eight-dimensional Hilbert space, $\{|000\rangle, |001\rangle, |010\rangle, |011\rangle, |100\rangle, |101\rangle, |110\rangle, |111\rangle\}$, and their dual-corresponding bras.

(c) Obtain the matrix representation of $\hat{U}_{\mathrm{Toffoli}}$ given the above eight basis states. It should be an 8×8 matrix.

Exercise 4.9 (Quantum circuit)

Figure 4.20 A quantum circuit. Here A is the input $|\psi\rangle$, B is the input $|0\rangle$, and C is the input $|0\rangle$; H represents the Hadamard gate.

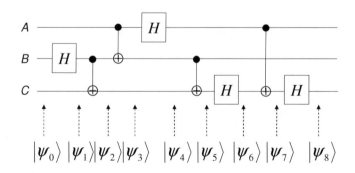

Let us start with the following state:

$$|\psi\rangle \otimes |0\rangle \otimes |0\rangle = (a|0\rangle + b|1\rangle) \otimes |0\rangle \otimes |0\rangle := |\psi 00\rangle, \qquad (4.65)$$

where $|a|^2 + |b|^2 = 1$. Consider what happens when we feed the product state $|\psi_0\rangle = |\psi 00\rangle$ into the quantum circuit shown in Figure 4.20. Examine what happens to this product state as it goes through the circuit, and find $|\psi_8\rangle$. Introducing the abbreviation $|i\rangle \otimes |j\rangle \otimes |k\rangle = |ijk\rangle$, state all intermediate results in the form

$$|\psi_i\rangle = a_0|000\rangle + a_1|001\rangle + a_2|010\rangle + a_3|011\rangle + a_4|100\rangle$$
$$+ a_5|101\rangle + a_6|110\rangle + a_7|111\rangle, \tag{4.66}$$

keeping the summands in this order.

5
Propagating Electron Waves in Engineered Potentials

Learning objectives:

- Learning how to use unbound states in analyzing particle dynamics as wave propagation.

- Understanding the phenomenon, implications, and applications of quantum tunneling.

- Applying the propagation matrix method to study various 1D potential problems.

- Becoming familiar with the basic concepts of the band theory of solids for electrons in a periodic potential.

THE CURRENT–VOLTAGE CHARACTERISTICS of modern electronic devices consisting of semiconductor heterostructures, such as resonant tunneling diodes, quantum cascade lasers, and tandem solar cells, are determined by the dynamics of electrons propagating through quantum-engineered 1D potential landscapes. In this chapter, we will develop a general formalism with which to describe transmission probabilities for electron waves propagating through arbitrary potentials, which can be used for analyzing electron motion in semiconductor devices. Furthermore, we will extend our formalism to 1D electrons moving in a general spatially periodic potential, based on which we will describe the basic concepts of the band theory of solids. The central theorem in band theory is the Bloch theorem, which we will derive and then use for discussing the dynamics of electrons in crystalline solids (or Bloch electrons).

5.1 Probability Current

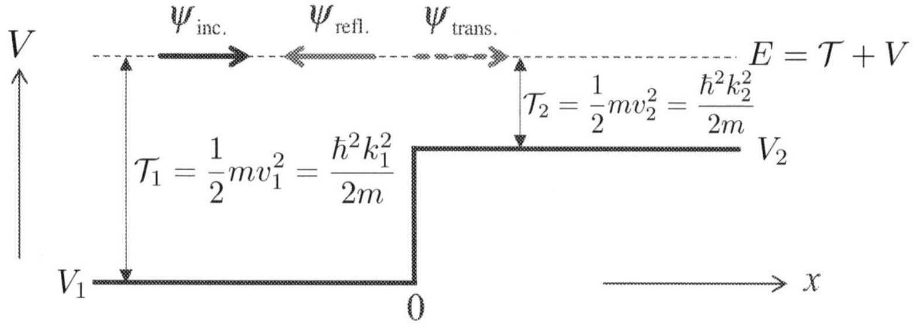

Figure 5.1 Step-function potential barrier. An electron with total energy E $(> V_2 > V_1)$ is incident from the left (Region 1: $x < 0$). We want to calculate the probability that the electron enters Region 2 ($x > 0$) as a function of E.

Most of the electronic states we encounter in this chapter are unbound states, which do not vanish at $x = \infty$ or $-\infty$. Let us consider the step-function potential barrier shown in Figure 5.1. The potential energy $V(x)$ is V_1 in Region 1 ($x < 0$) and V_2 in Region 2 ($x > 0$), where $V_2 > V_1$. An electron with total energy E is incident from the left (Region 1: $x < 0$). We want to calculate the transmission probability, P_t, that the electron transmits through the boundary to enter Region 2 ($x > 0$) as a function of E.[1]

A propagating state characteristically has a wavenumber given by $k = \sqrt{2m(E - V)}/\hbar$, which is real and positive since $E > V$. The corresponding kinetic energy $\mathcal{T} = \hbar^2 k^2/2m = E - V$ is positive. Since $V_2 > V_1$ in our problem, we immediately see that $\mathcal{T}_1 > \mathcal{T}_2$, which in turn means that $k_1 > k_2$. See Figure 5.1. Note that the energy, E, and thus, the frequency, $\omega = E/\hbar$, is conserved, but the wavelength, $\lambda = 2\pi/k$, changes when the electron moves from Region 1 to 2.[2]

The wavefunction in Region 1 is written as

$$\psi_1(x) = \psi_{\text{inc.}} + \psi_{\text{refl.}} = Ae^{ik_1 x} + Be^{-ik_1 x}, \tag{5.1}$$

whereas that in Region 2 is

$$\psi_2(x) = \psi_{\text{trans.}} = Ce^{ik_2 x}. \tag{5.2}$$

Thus, on the basis of what we learned in Section 2.3, the boundary conditions at $x = 0$ are

$$A + B = C, \tag{5.3}$$

$$k_1(A - B) = k_2 C, \tag{5.4}$$

from which we obtain

$$\left|\frac{B}{A}\right|^2 = \left(\frac{1 - k_2/k_1}{1 + k_2/k_1}\right)^2, \tag{5.5}$$

$$\left|\frac{C}{A}\right|^2 = \frac{4}{(1 + k_2/k_1)^2}. \tag{5.6}$$

Equation (5.6) is counterintuitive from a first look if we recall that $k_1 > k_2$. That is, it tells us that $|C|^2 > |A|^2$, which seems as though $P_t > 1$ (amplification?!). In analyzing the dynamics or flow of electrons represented by an unbound state, the probability current density, $j(x,t)$, is a more appropriate quantity to use than the probability density, $P(x,t)$, itself. They are connected through the following probability density conservation equation:[3]

$$\frac{\partial P(x,t)}{\partial t} = -\frac{\partial j(x,t)}{\partial x}. \tag{5.7}$$

[1] The classical mechanics results are simple and clear: $P_t = 1$ for $E > V_2$ and $P_t = 0$ for $V_1 < E \leq V_2$. However, quantum mechanically, we will find that $P_t < 1$ even for $E > V_2$, in general, although $P_t = 0$ for $V_1 < E \leq V_2$ as in the classical case.

[2] This is analogous to a light wave propagating from one region to another region with a different refractive index. See Appendix B.

[3] In terms of the charge density $\rho = eP = e|\psi|^2$ and the charge current density $J = ej$, Equation (5.7) can be written as $\partial\rho/\partial t = -\partial J/\partial x$, which is the charge conservation equation (also known as the continuity equation) familiar in electromagnetism. See Appendix C.

Using the time-dependent Schrödinger equation [Equation (2.79)] and its complex conjugate, we can derive

$$
\begin{aligned}
\frac{\partial P(x,t)}{\partial t} &= \frac{\partial |\psi(x,t)|^2}{\partial t} = \frac{\partial \psi^*}{\partial t}\psi + \psi^*\frac{\partial \psi}{\partial t} \\
&= \frac{1}{i\hbar}\frac{\hbar^2}{2m}\left(\frac{\partial^2 \psi^*}{\partial x^2}\psi - \psi^*\frac{\partial^2 \psi}{\partial x^2}\right) \\
&= -\frac{\partial}{\partial x}\left[\frac{\hbar}{2im}\left(\psi^*\frac{\partial \psi}{\partial x} - \frac{\partial \psi^*}{\partial x}\psi\right)\right],
\end{aligned}
\tag{5.8}
$$

which indicates that the probability current density is

$$
j(x,t) := \frac{\hbar}{2im}\left(\psi^*\frac{\partial \psi}{\partial x} - \frac{\partial \psi^*}{\partial x}\psi\right).
\tag{5.9}
$$

Using this definition, we can now calculate the probability currents for the incident, reflected, and transmitted waves in Figure 5.1 to be

$$
j_{\text{inc.}} = \frac{\hbar k_1}{m}|A|^2,
\tag{5.10}
$$

$$
j_{\text{refl.}} = -\frac{\hbar k_1}{m}|B|^2,
\tag{5.11}
$$

$$
j_{\text{trans.}} = \frac{\hbar k_2}{m}|C|^2.
\tag{5.12}
$$

The transmission probability P_t can now be calculated:

$$
P_t = \left|\frac{j_{\text{trans.}}}{j_{\text{inc.}}}\right| = \frac{k_2}{k_1}\left|\frac{C}{A}\right|^2.
\tag{5.13}
$$

Substituting Equation (5.6) into this equation, we find that

$$
P_t = \frac{4k_1 k_2}{(k_1 + k_2)^2}.
\tag{5.14}
$$

Similarly, the reflection probability is given by[4]

$$
P_r = \left|\frac{j_{\text{refl.}}}{j_{\text{inc.}}}\right| = \left|\frac{B}{A}\right|^2 = \left(\frac{1 - k_2/k_1}{1 + k_2/k_1}\right)^2 = \left(\frac{k_1 - k_2}{k_1 + k_2}\right)^2.
\tag{5.15}
$$

One can easily show that j can also be written as

$$
j(x,t) = \text{Re}\left(\psi^*\frac{\hbar}{im}\frac{\partial \psi}{\partial x}\right).
\tag{5.16}
$$

In this sense, the conservation of probability current across an interface at $x = x_0$, i.e., $j_1|_{x=x_0} = j_2|_{x=x_0}$, is guaranteed by the boundary conditions that we have been using:

$$
\psi_1|_{x=x_0} = \psi_2|_{x=x_0} \quad \text{and} \quad \left.\frac{\partial \psi_1}{\partial x}\right|_{x=x_0} = \left.\frac{\partial \psi_2}{\partial x}\right|_{x=x_0}.
\tag{5.17}
$$

[4] Note that $P_t + P_r = 1$, as expected.

However, in semiconductor heterostructures, different layers generally have different effective masses.[5] Therefore, in order to conserve the probability current density, the second boundary condition above has to be modified to

[5] See Equation (5.95).

$$\frac{1}{m_1} \frac{\partial \psi_1}{\partial x}\bigg|_{x=x_0} = \frac{1}{m_2} \frac{\partial \psi_2}{\partial x}\bigg|_{x=x_0}. \tag{5.18}$$

Using these generalized boundary conditions, we can calculate the reflection probability to be

$$P_r = \left(\frac{1 - m_1 k_2/m_2 k_1}{1 + m_1 k_2/m_2 k_1}\right)^2 = \left(\frac{m_2 k_1 - m_1 k_2}{m_2 k_1 + m_1 k_2}\right)^2. \tag{5.19}$$

An interesting consequence of this is that now P_r can be zero (and thus, P_t can be unity), when $m_2 k_1 = m_1 k_2$.[6]

[6] A vanishing P_r would be impossible to realize if $m_1 = m_2$ since $k_1 > k_2$. However, this can be achieved when $m_1 > m_2$.

5.2   Quantum Tunneling

Let us next consider the single square-potential barrier shown in Figure 5.2. The potential energy is V_0 (> 0) in Region 2 ($0 \leq x \leq L$) and zero otherwise. An electron with total energy E is incident from the left (Region 1: $x < 0$). Our goal is to calculate the transmission probability, P_t, that the electron transmits through the potential barrier to enter Region 3 ($x > L$), as a function of E. According to classical mechanics, $P_t = 1$ for $E > V_0$ and $P_t = 0$ for $0 < E \leq V_0$. However, we will find that, quantum mechanically, $P_t \leq 1$ even for $E > V_0$; furthermore, $P_t \neq 0$ even for $0 < E \leq V_0$, which is the phenomenon of tunneling.

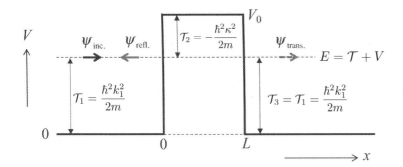

Figure 5.2 Electronic tunneling through a singe-barrier potential with a barrier height of V_0. An electron with total energy E is incident from the left (Region 1: $x < 0$). We want to calculate the probability that the electron enters Region 3 ($x > L$) as a function of E.

Let us start with the case in which $0 < E \leq V_0$, as depicted in Figure 5.2. In both Region 1 ($x < 0$) and Region 3 ($x > L$), the kinetic energy is $\mathcal{T}_1 = \mathcal{T}_3 = \hbar^2 k_1^2/2m$, where $k_1 = \sqrt{2mE}/\hbar$. However, in Region 2 ($0 \leq x \leq L$), the total energy E is *smaller* than the potential energy, and so *the kinetic energy is negative*, i.e., $\mathcal{T}_2 = -\hbar^2 \kappa^2/2m$, where $\kappa = \sqrt{2m(V_0 - E)}/\hbar$. Classically, no electron can exist in Region 2, as implied by the negative kinetic energy.

As the general solution to the time-independent Schrödinger equation, the wavefunction in Region 1 is written as

$$\psi_1(x) = \psi_{\text{inc.}} + \psi_{\text{refl.}} = Ae^{ik_1x} + Be^{-ik_1x}. \qquad (5.20)$$

Similarly, the wavefunction in Region 2 is given by

$$\psi_2(x) = Ce^{\kappa x} + De^{-\kappa x}, \qquad (5.21)$$

and that in Region 3 is

$$\psi_3(x) = \psi_{\text{trans.}} = Fe^{ik_1x}. \qquad (5.22)$$

We are interested in calculating the ratio, F/A, as the transmission probability is $P_t = |F/A|^2$. This can be obtained through boundary conditions, as follows. The boundary conditions [Equation (5.17)] at $x = 0$ and L read[7]

$$A + B = C + D, \qquad (5.23)$$

$$ik_1(A - B) = \kappa(C - D), \qquad (5.24)$$

$$Ce^{\kappa L} + De^{-\kappa L} = Fe^{ik_1L}, \qquad (5.25)$$

$$\kappa(Ce^{i\kappa L} - De^{-i\kappa L}) = ik_1Fe^{ik_1L}. \qquad (5.26)$$

By eliminating B, C, and D from these equations in a straightforward manner, we can derive

$$P_t = \left| \frac{F}{A} \right|^2 = \left[1 + \left(\frac{k_1^2 + \kappa^2}{2k_1\kappa} \right)^2 \sinh^2(\kappa L) \right]^{-1}. \qquad (5.27)$$

By using $k_1 = \sqrt{2mE}/\hbar$ and $\kappa = \sqrt{2m(V_0 - E)}/\hbar$, and introducing a dimensionless energy $\varepsilon := E/V_0$, Equation (5.27) can be rewritten as

$$P_t(0 \leq \varepsilon \leq 1) = \left[1 + \frac{\sinh^2(\eta\sqrt{1-\varepsilon})}{4\varepsilon(1-\varepsilon)} \right]^{-1}, \qquad (5.28)$$

where the parameter

$$\eta := \frac{\sqrt{2mV_0}}{\hbar}L \qquad (5.29)$$

is also dimensionless. The transmission probability P_t increases monotonically[8] with ε and reaches

$$P_t(\varepsilon = 1) = \frac{1}{1 + mL^2V_0/2\hbar^2} = \left(1 + \frac{\eta^2}{4} \right)^{-1}, \qquad (5.30)$$

[7] It is assumed that the mass m of the electron is constant. See Exercise 5.6 for the case where the mass is not constant.

[8] When $\varepsilon \sim 0$, $P_t \propto \varepsilon^2$.

when $\varepsilon = 1$. As expected, this value approaches zero as $L \to \infty$.

In the limit where the transmission probability is small, $P_t \ll 1$, one can assume that $\kappa L = \eta\sqrt{1-\varepsilon} \gg 1$.[9] Since $\sinh(\kappa L) \approx \frac{1}{2}\kappa L$ in this limit and the second term dominates inside the bracket in Equation (5.28), one can approximate as follows:

$$P_t \approx \left[\frac{\frac{1}{4}\exp^2(\eta\sqrt{1-\varepsilon})}{4\varepsilon(1-\varepsilon)}\right]^{-1} = 16\varepsilon(1-\varepsilon)e^{-2\eta\sqrt{1-\varepsilon}}$$

$$= \frac{16E(V_0-E)}{V_0^2}e^{-2\eta\sqrt{1-\varepsilon}} = P_{t0}\,e^{-2\kappa L}, \tag{5.31}$$

where $P_{t0} = 16E(V_0 - E)/V_0^2 = 16\varepsilon(1-\varepsilon)$.

[9] This happens when V_0 and/or L is large, i.e., when the barrier is high and/or wide.

Example 5.1 Quantum tunneling and STM

Let us consider an electron with $E = 2$ eV incident on a potential barrier with $V_0 = 20$ eV and $L = 1$ nm. We can then calculate the following: $\varepsilon = 2/20 = 0.1$, $\eta = \sqrt{2 \times 9.11 \times 10^{-31} \times 20 \times 1.6 \times 10^{-19}} \times 1 \times 10^{-9}/1.05 \times 10^{-34} = 23.0$, $\kappa L = \eta\sqrt{1-\varepsilon} = 21.82$, and $P_{t0} = 16 \times 0.1 \times 0.9 = 1.44$. If we use the full expression [Equation (5.28)], we get $P_t = 1.61 \times 10^{-19}$. This is a very small number, but it is not zero. If we use the expression appropriate for the low transmission probability regime [Equation (5.31)], we obtain $P_t = 1.44 \times \exp(-2 \times 21.82) = 1.12 \times 10^{-19}$.

If we decrease the barrier width from 1 nm to 0.1 nm, then we see that $\eta = 2.30$, $\kappa L = 2.18$, and $P_t = 1.85 \times 10^{-2}$. Namely, reducing the tunneling barrier width by one order of magnitude results in an increase in transmission probability by 17 orders of magnitude! This ultrahigh sensitivity of the transmission probability to distance is the fundamental idea behind scanning tunneling microscopy (STM), used for imaging material surfaces with atomic resolution.[10] Figure 5.3 shows an STM setup schematically.

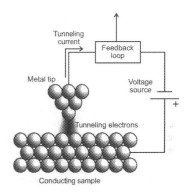

Figure 5.3 A schematic diagram of a scanning tunneling microscopy setup. See, e.g., C. Julian Chen, *Introduction to Scanning Tunneling Microscopy, Second Edition* (Oxford University Press, 1993) for more details about STM. © Deyin Kong (Rice University).

[10] STM was invented by Gerd Binnig and Heinrich Rohrer at IBM Zürich, and they were awarded the 1986 Nobel Prize in Physics.

When $E > V_0$, we can calculate the transmission probability by replacing κ with $ik_2 = i\sqrt{2m(E-V_0)}/\hbar$ in Equation (5.27) as

$$P_t = \left|\frac{F}{A}\right|^2 = \left[1 + \left(\frac{k_1^2 - k_2^2}{2k_1 k_2}\right)^2 \sin^2(k_2 L)\right]^{-1}. \tag{5.32}$$

Alternatively, using the dimensionless energy ε, one can write

$$P_t(\varepsilon \geq 1) = \left[1 + \frac{\sin^2(\eta\sqrt{\varepsilon-1})}{4\varepsilon(\varepsilon-1)}\right]^{-1}. \tag{5.33}$$

Note that, because of the $\sin^2(k_2 L)$ factor, P_t *oscillates* as a function of energy but asymptotically approaches 1 as $E \to \infty$. However, when $k_2 L = n\pi$, where $n = 1, 2, 3, \ldots$, $P_t = 1$ (and thus, $P_r = 0$).

[11] This is the Fabry–Pérot resonance condition, commonly observed for an optical wave transmitting through a thin film. See Exercise 6.2.

At these special values of $k_2 L$, the corresponding wavelength $\lambda_2 = 2\pi/k_2$ satisfies $n(\lambda_2/2) = L$.[11]   Figure 5.4 plots the transmission probability as a function of dimensionless energy when $\eta = 7$.

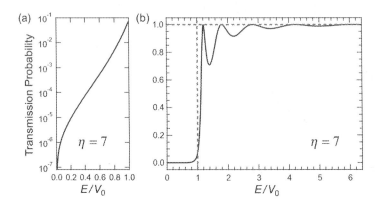

Figure 5.4 Transmission probability as a function of incident energy normalized by the barrier height, E/V_0, through the tunneling barrier shown in Figure 5.2 when $\eta = 7$. Part (a) is a close-up of the start of the curve shown in (b).

5.3   Propagation Matrix Method

While working on the step-function barrier and single-barrier tunneling problems in the previous sections, we noticed that the number of coupled equations to solve increases as the number of interfaces (or potential steps) increases – each interface introduces two equations coming from the continuity of the wavefunction, ψ, and the first derivative of the wavefunction, $d\psi/dx$. Thus, the problem can become unmanageably complicated once we start looking at real semiconductor heterostructure devices, which typically consist of tens or hundreds of different layers and interfaces.

To address this difficulty, here we introduce the propagation matrix method, which significantly reduces the mathematical complexity in describing the propagation of electron waves in 1D potentials. The method decomposes the given potential into a series of steps (representing heterointerfaces) and flat regions between steps (representing layers). We associate each step with a 2×2 matrix, describing the transmission and reflection coefficients at the interface; we associate each flat region with a 2×2 matrix describing the "free" propagation of the electron wave in that layer. Finally, the amplitude of the wave after its transmission through the potential is connected to the amplitude of the incident wave through the 2×2 matrix that is the product of all the intermediate matrices.

5.3.1 Single-Barrier Problem

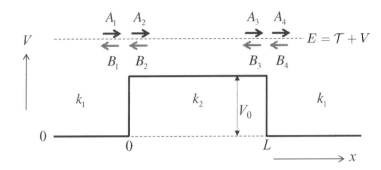

Figure 5.5 Single-barrier problem solved in the formalism of the propagation matrix method. $k_j = \sqrt{2m(E - V_j)}/\hbar$, where $j = 1, 2$.

Let us start with the first interface, i.e., $x = 0$, in Figure 5.5. On the left side of the interface, the wavefunction is $\psi_1 = A_1 e^{ik_1 x} + B_1 e^{-ik_1 x}$, whereas, on the right side, $\psi_2 = A_2 e^{ik_2 x} + B_2 e^{-ik_2 x}$. By connecting the two wavefunctions smoothly at $x = 0$, we get the following two coupled equations for A_1, B_1, A_2, and B_2:

$$A_1 + B_1 = A_2 + B_2, \tag{5.34}$$

$$k_1(A_1 - B_1) = k_2(A_2 - B_2). \tag{5.35}$$

We can rewrite these equations in matrix form as:

$$\begin{bmatrix} A_1 \\ B_1 \end{bmatrix} = \begin{bmatrix} 1 & 1 \\ k_1 & -k_1 \end{bmatrix}^{-1} \begin{bmatrix} 1 & 1 \\ k_2 & -k_2 \end{bmatrix} \begin{bmatrix} A_2 \\ B_2 \end{bmatrix}$$

$$= \frac{1}{2} \begin{bmatrix} 1 + \dfrac{k_2}{k_1} & 1 - \dfrac{k_2}{k_1} \\ 1 - \dfrac{k_2}{k_1} & 1 + \dfrac{k_2}{k_1} \end{bmatrix} \begin{bmatrix} A_2 \\ B_2 \end{bmatrix} := \mathbf{P}_1 \begin{bmatrix} A_2 \\ B_2 \end{bmatrix}. \tag{5.36}$$

Next let us consider the propagation of the electron wave through the $0 < x < L$ region. In this region, the electron wave propagates freely, acquiring a phase factor that is proportional to the distance of propagation. Namely, $A_3 = A_2 e^{ik_2 L}$ and $B_2 = B_3 e^{ik_2 L}$, which can be written in matrix form as:

$$\begin{bmatrix} A_2 \\ B_2 \end{bmatrix} = \begin{bmatrix} e^{-ik_2 L} & 0 \\ 0 & e^{ik_2 L} \end{bmatrix} \begin{bmatrix} A_3 \\ B_3 \end{bmatrix} := \mathbf{P}_2 \begin{bmatrix} A_3 \\ B_3 \end{bmatrix}. \tag{5.37}$$

Finally, let us consider the second interface, located at $x = L$. The situation is reversed, as compared with the situation at $x = 0$, and from symmetry, we can relate (A_3, B_3) and (A_4, B_4) as

$$\begin{bmatrix} A_3 \\ B_3 \end{bmatrix} = \frac{1}{2} \begin{bmatrix} 1 + \dfrac{k_1}{k_2} & 1 - \dfrac{k_1}{k_2} \\ 1 - \dfrac{k_1}{k_2} & 1 + \dfrac{k_1}{k_2} \end{bmatrix} \begin{bmatrix} A_4 \\ B_4 \end{bmatrix} = \mathbf{P}_3 \begin{bmatrix} A_4 \\ B_4 \end{bmatrix}. \tag{5.38}$$

By assembling the above equations involving $\mathbf{P}_1$, $\mathbf{P}_2$, and $\mathbf{P}_3$, we can directly relate (A_1, B_1) and (A_4, B_4):

$$\begin{bmatrix} A_1 \\ B_1 \end{bmatrix} = \mathbf{P}_1 \begin{bmatrix} A_2 \\ B_2 \end{bmatrix} = \mathbf{P}_1\mathbf{P}_2 \begin{bmatrix} A_3 \\ B_3 \end{bmatrix} = \mathbf{P}_1\mathbf{P}_2\mathbf{P}_3 \begin{bmatrix} A_4 \\ B_4 \end{bmatrix} := \mathbf{P} \begin{bmatrix} A_4 \\ B_4 \end{bmatrix}, \quad (5.39)$$

where

$$\mathbf{P} = \mathbf{P}_1\mathbf{P}_2\mathbf{P}_3 = \begin{bmatrix} P_{11} & P_{12} \\ P_{21} & P_{22} \end{bmatrix}$$

$$= \frac{1}{2} \begin{bmatrix} 1 + \dfrac{k_2}{k_1} & 1 - \dfrac{k_2}{k_1} \\ 1 - \dfrac{k_2}{k_1} & 1 + \dfrac{k_2}{k_1} \end{bmatrix} \begin{bmatrix} e^{-ik_2 L} & 0 \\ 0 & e^{ik_2 L} \end{bmatrix} \frac{1}{2} \begin{bmatrix} 1 + \dfrac{k_1}{k_2} & 1 - \dfrac{k_1}{k_2} \\ 1 - \dfrac{k_1}{k_2} & 1 + \dfrac{k_1}{k_2} \end{bmatrix}, $$

$$(5.40)$$

and

$$P_{11} = \frac{1}{4} \left\{ e^{-ik_2 L} \left(1 + \frac{k_2}{k_1}\right) \left(1 + \frac{k_1}{k_2}\right) + e^{ik_2 L} \left(1 - \frac{k_2}{k_1}\right) \left(1 - \frac{k_1}{k_2}\right) \right\},$$

$$(5.41)$$

$$P_{12} = \frac{1}{4} \left\{ e^{ik_2 L} \left(1 - \frac{k_2}{k_1}\right) \left(1 + \frac{k_1}{k_2}\right) + e^{-ik_2 L} \left(1 + \frac{k_2}{k_1}\right) \left(1 - \frac{k_1}{k_2}\right) \right\},$$

$$(5.42)$$

$$P_{21} = \frac{1}{4} \left\{ e^{-ik_2 L} \left(1 - \frac{k_2}{k_1}\right) \left(1 + \frac{k_1}{k_2}\right) + e^{ik_2 L} \left(1 + \frac{k_2}{k_1}\right) \left(1 - \frac{k_1}{k_2}\right) \right\},$$

$$(5.43)$$

$$P_{22} = \frac{1}{4} \left\{ e^{ik_2 L} \left(1 + \frac{k_2}{k_1}\right) \left(1 + \frac{k_1}{k_2}\right) + e^{-ik_2 L} \left(1 - \frac{k_2}{k_1}\right) \left(1 - \frac{k_1}{k_2}\right) \right\}.$$

$$(5.44)$$

Since there is no component propagating to the left on the right of the barrier, $B_4 = 0$ in the present case. Therefore,

$$\begin{bmatrix} A_1 \\ B_1 \end{bmatrix} = \begin{bmatrix} P_{11} & P_{12} \\ P_{21} & P_{22} \end{bmatrix} \begin{bmatrix} A_4 \\ 0 \end{bmatrix} = \begin{bmatrix} P_{11} A_4 \\ P_{21} A_4 \end{bmatrix}, \quad (5.45)$$

i.e.,

$$P_{\mathrm{t}} = \left| \frac{A_4}{A_1} \right|^2 = \frac{1}{|P_{11}|^2}. \quad (5.46)$$

We can rewrite Equation (5.41) as

$$P_{11} = \frac{1}{4} \left[2(e^{ik_2 L} + e^{-ik_2 L}) - \frac{k_1^2 + k_2^2}{k_1 k_2} (e^{ik_2 L} - e^{-ik_2 L}) \right]$$

$$= \cos(k_2 L) - i \left(\frac{k_1^2 + k_2^2}{2k_1 k_2} \right) \sin(k_2 L). \quad (5.47)$$

Thus,

$$|P_{11}|^2 = \cos^2(k_2 L) + \left(\frac{k_1^2 + k_2^2}{2k_1 k_2}\right)^2 \sin^2(k_2 L)$$

$$= 1 + \left(\frac{k_1^2 - k_2^2}{2k_1 k_2}\right)^2 \sin^2(k_2 L), \tag{5.48}$$

and

$$P_t^{E>V_0} = \frac{1}{|P_{11}|^2} = \left[1 + \left(\frac{k_1^2 - k_2^2}{2k_1 k_2}\right)^2 \sin^2(k_2 L)\right]^{-1}, \tag{5.49}$$

which agrees with Equation (5.32).

5.3.2 Properties of Propagation Matrices

The components of a propagation matrix are related to each other through time-reversal symmetry, i.e., because the Hamiltonian does not change under the replacement of k with $-k$.

Let us consider the propagation matrix representing the interface between the jth and $(j+1)$th layers:

$$\begin{bmatrix} A_j \\ B_j \end{bmatrix} = \mathbf{P} \begin{bmatrix} A_{j+1} \\ B_{j+1} \end{bmatrix} = \begin{bmatrix} P_{11} & P_{12} \\ P_{21} & P_{22} \end{bmatrix} \begin{bmatrix} A_{j+1} \\ B_{j+1} \end{bmatrix}. \tag{5.50}$$

The wavefunction in the jth layer, $\psi_j(x) = A_j e^{ik_j x} + B_j e^{-ik_j x}$, satisfies the Schrödinger equation

$$\left(-\frac{\hbar^2}{2m}\frac{d^2}{dx^2} + V_j\right)\psi_j = E\psi_j. \tag{5.51}$$

By taking the complex conjugate of this equation, we get

$$\left(-\frac{\hbar^2}{2m}\frac{d^2}{dx^2} + V_j\right)\psi_j^* = E\psi_j^*. \tag{5.52}$$

Thus, $\psi_j^*(x)$ is also a solution to the same Schrödinger equation. However, since $\psi_j^*(x) = A_j^* e^{-ik_j x} + B_j^* e^{ik_j x}$, we see that the roles of the coefficients are reversed. Namely, B_j^* represents the amplitude of the wave propagating to the right, i.e., the incident wave, whereas A_j^* now represents the reflected component. Therefore, we can relate (B_j^*, A_j^*) and (B_{j+1}^*, A_{j+1}^*) using the same propagation matrix that relates (A_j, B_j) and (A_{j+1}, B_{j+1}), i.e.,

$$\begin{bmatrix} B_j^* \\ A_j^* \end{bmatrix} = \begin{bmatrix} P_{11} & P_{12} \\ P_{21} & P_{22} \end{bmatrix} \begin{bmatrix} B_{j+1}^* \\ A_{j+1}^* \end{bmatrix}. \tag{5.53}$$

By taking the complex conjugate of Equation (5.53), one finds

$$\begin{bmatrix} B_j \\ A_j \end{bmatrix} = \begin{bmatrix} P_{11}^* & P_{12}^* \\ P_{21}^* & P_{22}^* \end{bmatrix} \begin{bmatrix} B_{j+1} \\ A_{j+1} \end{bmatrix}. \tag{5.54}$$

Furthermore, by rearranging rows and columns, one obtains

$$\begin{bmatrix} A_j \\ B_j \end{bmatrix} = \begin{bmatrix} P_{22}^* & P_{21}^* \\ P_{12}^* & P_{11}^* \end{bmatrix} \begin{bmatrix} A_{j+1} \\ B_{j+1} \end{bmatrix}. \tag{5.55}$$

[12] See another property of propagation matrices shown in Exercise 5.10.

Then, comparing Equations (5.50) and (5.55), we can conclude that[12]

$$P_{11}^* = P_{22}, \tag{5.56}$$

$$P_{21}^* = P_{12}. \tag{5.57}$$

5.3.3   Double-Barrier Tunneling

Next we investigate the tunneling phenomenon in the double-barrier potential shown in Figure 5.6.

Figure 5.6 Double-barrier potential, consisting of two identical potential barriers with height V_0 and width L separated by a distance W.

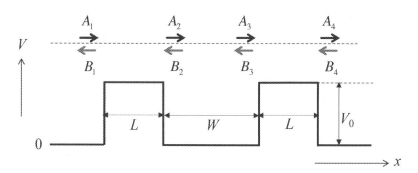

One would basically have to set up boundary conditions at each interface, which in this case would lead to eight coupled equations. However, we can greatly reduce the mathematical complexity of this problem using the propagation matrix method, because we have full knowledge for the single-barrier case. Let us define the propagation matrix for the single-barrier case as[13]

[13] $P_{SB,11}$ and $P_{SB,21}$ are given by Equations (5.41) and (5.43), respectively.

$$\mathbf{P}_{SB} = \begin{bmatrix} P_{SB,11} & P_{SB,21}^* \\ P_{SB,21} & P_{SB,11}^* \end{bmatrix}. \tag{5.58}$$

Using $\mathbf{P}_{SB}$, we can write

$$\begin{bmatrix} A_1 \\ B_1 \end{bmatrix} = \mathbf{P}_{SB} \begin{bmatrix} A_2 \\ B_2 \end{bmatrix}, \tag{5.59}$$

$$\begin{bmatrix} A_3 \\ B_3 \end{bmatrix} = \mathbf{P}_{SB} \begin{bmatrix} A_4 \\ B_4 \end{bmatrix}. \tag{5.60}$$

In addition, the propagation in the middle region can be described by

$$\begin{bmatrix} A_2 \\ B_2 \end{bmatrix} = \begin{bmatrix} e^{-ik_1 W} & 0 \\ 0 & e^{ik_1 W} \end{bmatrix} \begin{bmatrix} A_3 \\ B_3 \end{bmatrix}. \tag{5.61}$$

Combining these relations, we can relate (A_1, B_1) and (A_4, B_4):

$$\begin{bmatrix} A_1 \\ B_1 \end{bmatrix} = \mathbf{P}_{SB} \begin{bmatrix} e^{-ik_1 W} & 0 \\ 0 & e^{ik_1 W} \end{bmatrix} \mathbf{P}_{SB} \begin{bmatrix} A_4 \\ B_4 \end{bmatrix} = \mathbf{P}_{DB} \begin{bmatrix} A_4 \\ B_4 \end{bmatrix}$$

$$= \begin{bmatrix} e^{-ik_1 W} P_{SB,11}^2 + e^{ik_1 W} |P_{SB,21}|^2 & P_{DB,21}^* \\ e^{-ik_1 W} P_{SB,11} P_{SB,21} + e^{ik_1 W} P_{SB,11}^* P_{SB,21}^* & P_{DB,11}^* \end{bmatrix} \begin{bmatrix} A_4 \\ B_4 \end{bmatrix}. \tag{5.62}$$

Since $B_4 = 0$ as before, the transmission probability is given by

$$P_t = \left| \frac{A_4}{A_1} \right|^2 = \frac{1}{|P_{DB,11}|^2}. \tag{5.63}$$

5.3.4 Propagation Matrix for a δ-Function Potential

Next, let us examine the propagation matrix for a δ-function potential barrier. We will use this result when we consider the dynamics of an electron traveling in a periodic potential such as that found in a solid state crystal. There are many ways to get to a δ-function potential; we will model ours as a rectangular potential barrier whose width $L \to 0$ while the potential step $V_0 \to \infty$ in such a way that the product $V_0 L$ is held constant. This is depicted in Figure 5.7.

As with the rectangular potential barrier considered in Section 5.3.1, the electron has a wavevector $k_1 = \sqrt{2mE}/\hbar$ in the regions to the left and right of the potential barrier. However, in the barrier region ($0 \le x \le L$), the potential V_0 is greater than the total energy E, and hence, the kinetic energy is negative, and the wavevector $k_2 = \sqrt{2m(E - V_0)}/\hbar$ is imaginary. Since $V_0 \gg E$, we can approximate k_2 as $k_2 \approx \sqrt{2m(-V_0)}/\hbar$.

We know P_{11} and P_{12} for the rectangular potential barrier from Section 5.3.1, but we need to take into account the fact that the wavevector k_2 in the barrier region is imaginary. So, we make the substitution $k_2 \to ik_2$, recognizing that there are now exponentials instead of plane waves inside the barrier. Everything else stays the same as before. With this substitution, we get from Equation (5.41)

$$P_{11} = \frac{-(k_1^2 - k_2^2)}{4ik_1 k_2} (e^{-k_2 L} - e^{k_2 L}) + \frac{1}{2}(e^{-k_2 L} - e^{k_2 L}). \tag{5.64}$$

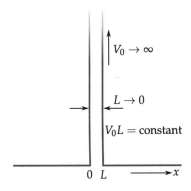

Figure 5.7 A δ-function potential barrier seen as the limit of a rectangular potential. The strength of the potential is characterized by $V_0 L =$ constant. © Deyin Kong (Rice University).

Note that since $ik_2 \propto \sqrt{V_0}$, in the limit $V_0 \to \infty$ and $L \to 0$, we have $k_2 L \to 0$. We can therefore expand $e^{k_2 L} \approx 1 + k_2 L$ and $e^{-k_2 L} \approx 1 - k_2 L$. With these approximations, we get

$$P_{11} = 1 + \frac{imV_0 L}{\hbar^2 k_1}. \tag{5.65}$$

Defining

$$q_0 := \frac{mV_0 L}{\hbar^2} \tag{5.66}$$

as the constant that quantifies the strength of the potential barrier $V_0 L$, we can rewrite P_{11} as

$$P_{11} = 1 + \frac{iq_0}{k_1}. \tag{5.67}$$

The transmission coefficient is now straightforward to evaluate, and is given by

$$P_t = \left| \frac{1}{P_{11}} \right|^2 = \left| \frac{1}{1 + iq_0/k_1} \right|^2 = \frac{k_1^2}{k_1^2 + q_0^2} = \frac{E}{E + \dfrac{\hbar^2 q_0^2}{2m}}. \tag{5.68}$$

The other matrix element we need is P_{12}. We know from Equation (5.42) that it is given by

$$P_{12} = \frac{k_1^2 - k_2^2}{4k_1 k_2} \left(e^{ik_2 L} - e^{-ik_2 L} \right), \tag{5.69}$$

which, under the substitutions $k_2 \to ik_2$ and $k_2 L \to 0$ as before, reduces to

$$P_{12} = i\frac{q_0}{k_1}. \tag{5.70}$$

Since $P_{22} = P_{11}^*$ and $P_{21} = P_{12}^*$, we can now write down the complete propagation matrix for the delta function potential as

$$\mathbf{P}_\delta = \begin{bmatrix} 1 + i\dfrac{q_0}{k_1} & i\dfrac{q_0}{k_1} \\[2ex] -i\dfrac{q_0}{k_1} & 1 - i\dfrac{q_0}{k_1} \end{bmatrix}. \tag{5.71}$$

5.4   Periodic Potentials and Band Theory

One of the most successful applications of quantum mechanics is in the explanation of the behavior of electrons in crystalline solids. Owing to the interaction with an array of atoms or ions (called a periodic lattice, schematically depicted in Figure 5.8), the electron states

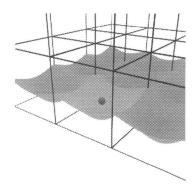

Figure 5.8 An electron in a crystal experiences a periodic potential energy. © Deyin Kong (Rice University).

and dynamics are significantly modified compared with those in free space. Proper incorporation of the lattice periodic potential into the Schrödinger equation leads to the formation of bands and band gaps. The resulting band structure is highly material-dependent and explains why different materials possess diverse electronic and optical properties, as detailed in Chapter 7.[14] In this section, we examine the band theory of solids by treating the lattice as a 1D periodic potential, using the methods we developed in the previous sections.

5.4.1 Bloch's Theorem

Bloch's theorem (known in mathematics as Floquet's theorem[15].) is of great importance in solid-state physics, as it represents an intuitive form for the electronic wavefunctions in a crystal lattice. It is essentially a prescription for writing down a valid form of the wavefunction for a particle that is subject to a periodic potential. Consider a 1D potential V such that $V(x + na) = V(x)$, where n is an integer. This defines a potential that is periodic in x with period a.

Bloch's theorem states that we can choose the eigenstates $\psi_k(x)$ of a one-electron Hamiltonian $\hat{H} = -(\hbar^2/2m)d^2/dx^2 + V(x)$, where $V(x + na) = V(x)$, to be a plane wave e^{ikx} times a function $u_k(x)$ that has the same periodicity as $V(x)$. That is, we can write the wavefunction as a Bloch function of the form

$$\psi_k(x) = u_k(x)\, e^{ikx}, \tag{5.72}$$

where $u_k(x + na) = u_k(x)$, and k is called the Bloch wavevector or Bloch index. Using the above Bloch form for the wavefunction, we can write the wavefunction at position $x + a$ as

$$\psi_k(x + a) = u_k(x + a)\, e^{ik(x+a)}$$
$$= u_k(x)\, e^{ikx}\, e^{ika} = \psi_k(x)\, e^{ika}. \tag{5.73}$$

Equations (5.72) and (5.73) can be regarded as equivalent statements of Bloch's theorem. Note that the probability density of finding the electron at position x is given by $|\psi_k(x)|^2 = |u_k(x)|^2$, and hence, it is modulated at the same period as the potential $V(x)$, as one would intuitively expect. The Bloch phase factor e^{ikx} is just a plane-wave-like phase that connects the wavefunctions of neighboring unit cells (repeat cells of the lattice).

All relevant information is carried by values of the Bloch wavevector k that lie within the so-called first Brillouin zone (FBZ), which is defined by $k \in [-\pi/a, \pi/a]$, where a is the spacing period.[16] To see this, consider a wavevector $k' = k + 2\pi n/a$. Then, $e^{ik'a} = e^{i(k+2\pi n/a)a} = e^{ika}e^{i\pi n} = e^{ika}$.

[14] For a more in-depth study of solid state physics, see, e.g., C. Kittel, *Introduction to Solid State Physics, Eighth Edition* (John Wiley & Sons, 2004) and S. H. Simon, *The Oxford Solid State Basics* (Oxford University Press, 2013).

[15] Achille Marie Gaston Floquet (1847–1920) was a French mathematician. Floquet theory is important for dynamical systems that are driven by a time-periodic field, including matter strongly driven by a laser field. See, e.g., J. H. Shirley, "Solution of the Schrödinger equation with a Hamiltonian periodic in time," *Physical Review* **138**, B979 (1965)

[16] Also known as the lattice constant or the unit cell length.

As a final remark, Bloch's theorem is applicable to *any* particle that sees a periodic Hamiltonian (not just an electron), and furthermore, it makes no assumptions about the shape or the strength of the potential as long as it is periodic in space.

5.4.2 Band Gaps and Band Dispersions

Armed with Bloch's theorem and the propagation matrix for the δ-function potential, we are now all set to tackle the problem of electron propagation in a periodic Coulombic potential created by the lattice atoms in a crystal. We will not solve for the Coulombic potential; instead, we will approximate it as a series of δ-function potentials. This approximation is known as the Kronig–Penney model (Figure 5.9). The spacing between the δ-functions is the lattice constant a. Using the ideas of Section 5.3, we denote the amplitudes of the forward- and backward-going waves just to the left of the repeat cell (at $x = 0^-$) as A and B, respectively, while C and D represent similar amplitudes one lattice spacing away at $x = a^-$ (see Figure 5.9).

Using the propagation matrix formalism, we can connect (A, B) with (C, D) as follows:

$$\begin{bmatrix} A \\ B \end{bmatrix} = \mathbf{P} \begin{bmatrix} C \\ D \end{bmatrix}, \tag{5.74}$$

where $\mathbf{P}$ is a propagation matrix that represents transmission through the δ-function barrier, followed by free propagation through a region of length a. Using Equations (5.71) and (5.37), we can thus write

$$\mathbf{P} = \mathbf{P}_\delta \, \mathbf{P}_{\text{free}} = \begin{bmatrix} 1 + i\dfrac{q_0}{k_1} & i\dfrac{q_0}{k_1} \\ -i\dfrac{q_0}{k_1} & 1 - i\dfrac{q_0}{k_1} \end{bmatrix} \begin{bmatrix} e^{-ik_1 a} & 0 \\ 0 & e^{ik_1 a} \end{bmatrix}$$

$$= \begin{bmatrix} \left(1 + i\dfrac{q_0}{k_1}\right) e^{-ik_1 a} & i\dfrac{q_0}{k_1} e^{ik_1 a} \\ P_{12}^* & P_{11}^* \end{bmatrix}, \tag{5.75}$$

where, as a reminder, $k_1 = \sqrt{2mE}/\hbar$ determines the energy of the electron and q_0 is a constant quantifying the strength of the δ-function potential; see Equation (5.66).

On the other hand, Bloch's theorem [Equation (5.73)] tells us that the wavefunction at $x = a^-$ is related to the wavefunction at $x = 0^-$ by a phase e^{ika} (k being the Bloch wavevector), that is,

$$\begin{bmatrix} A \\ B \end{bmatrix} = e^{-ika} \begin{bmatrix} C \\ D \end{bmatrix}. \tag{5.76}$$

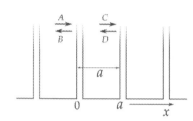

Figure 5.9 Schematic representation of the Kronig–Penney model as a series of periodic δ-function potentials. © Deyin Kong (Rice University).

Combining Equations (5.74), (5.75), and (5.76), we get

$$\begin{bmatrix} P_{11} - e^{-ika} & P_{12} \\ P_{21} & P_{22} - e^{-ika} \end{bmatrix} \begin{bmatrix} C \\ D \end{bmatrix} = 0. \qquad (5.77)$$

This corresponds to a homogeneous system of linear equations, which has a nontrivial solution only if the determinant of the 2×2 coefficient matrix is zero. This condition gives

$$(P_{11} - e^{-ika})(P_{22} - e^{-ika}) - P_{21}P_{12} = 0, \qquad (5.78)$$

$$P_{11}P_{22} + e^{-2ika} - P_{11}e^{-ika} - P_{22}e^{-ika} - P_{21}P_{12} = 0. \qquad (5.79)$$

Since $P_{11}P_{22} - P_{21}P_{12} \equiv \det \mathbf{P} = 1,$[17] this reduces to

$$e^{-2ika} - 2\,\mathrm{Re}(P_{11})e^{-ika} = -1. \qquad (5.80)$$

[17] See Exercise 5.10.

Taking the imaginary part of the above equation gives us the constraint

$$\mathrm{Re}(P_{11}) = \cos(ka). \qquad (5.81)$$

Substituting for $\mathrm{Re}(P_{11})$ above, we get

$$\cos(k_1 a) + q_0 a \left[\frac{\sin(k_1 a)}{k_1 a} \right] = \cos(ka). \qquad (5.82)$$

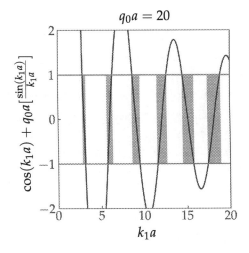

Figure 5.10 $\mathrm{Re}(P_{11})$ plotted as a function of $k_1 a$ for the Kronig–Penney model with $q_0 a = 20$. The allowed values of $k_1 a$ that result in a real Bloch wavevector k correspond to the gray areas between levels 1 and -1. The forbidden bands are the white areas between the gray areas. © Deyin Kong (Rice University).

Since the Bloch wavevector k is real, the expression on the left-hand side of Equation (5.82) must lie within $[-1, 1]$. To plot this expression graphically, we need to choose a value for $q_0 a$ (the barrier strength). The result is shown in Figure 5.10 for $q_0 a = 20$. We see that there are *allowed* and *forbidden* bands of k_1 that arise as a result of subjecting the electron to a periodic potential. Since $k_1 \propto \sqrt{E}$, this

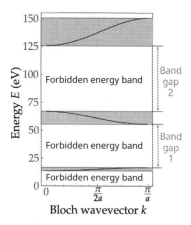

Figure 5.11 Energy band dispersion, $E = E(k)$, obtained for the Kronig–Penney model shown in Figure 5.9. The allowed values of $E = \hbar^2 k_1^2 / 2m$ are plotted as a function of the Bloch wavevector k within the first Brillouin zone, forming energy bands shaded in gray. The strength of the potential $q_0 a$ has been set to 20, with lattice constant $a = 0.15\,\text{nm}$. © Deyin Kong (Rice University).

[18] See Appendix B.

[19] Here, n is the band index and k is the Bloch wavevector (or, also known as the Bloch index).

is equivalent to saying that only those bands of energy are allowed which correspond to k_1 values in the shaded gray areas of Figure 5.10. To see this even more clearly, we need to plot the dispersion relation, that is, a relation that connects the Bloch wavevector k to the energy E of the electron. See the curves in Figure 5.11.

5.5 Dynamics of Bloch Electrons

The presence of a periodic potential drastically changes the way in which an electron moves. In this section, let us examine the dynamics of an electron in a crystal (or a Bloch electron) in an external force.

Recall that for any kind of wave (sound, light, electrons, …),[18]

$$\text{phase velocity } v_p = \omega / k, \text{ and} \tag{5.83}$$

$$\text{group velocity } v_g = \partial \omega / \partial k. \tag{5.84}$$

For a Bloch electron, there are bands of allowed energy described by the dispersion relation $\omega_n(k) = E_n(k)/\hbar$.[19] Thus, each Bloch state ψ_{nk} has a well-defined group velocity

$$\boxed{v_n(k) = \frac{\partial}{\partial k}\omega_n(k) = \frac{1}{\hbar}\frac{\partial}{\partial k}E_n(k)}. \tag{5.85}$$

There are also forbidden ranges in energy, called *band gaps*, where the electron cannot propagate freely. Clearly, this behavior is different from that of an electron in free space.

Finding the influence of a periodic potential $V(x)$ on electron motion requires the introduction of the concept of crystal momentum. Consider an external electric field, $\mathcal{E}$, that is applied to the crystal in the x-direction. As a result of this field, an electron experiences a force $-e\mathcal{E}$, causing a change in the electron's momentum. Let us take a closer look at *which* momentum it is that changes as a result of the applied external field. The total Hamiltonian for the electron can be written as

$$\hat{H} = \frac{\hat{p}_x^2}{2m} + V(x) - e|\mathcal{E}|x. \tag{5.86}$$

This Hamiltonian must satisfy the time-dependent Schrödinger equation, Equation (2.79). Suppose that, at time $t = 0$, $\psi(x,t)$ is equal to $\psi(x,0)$. Then, the time evolution[20] of the wavefunction can be formally written as

$$\psi(x,t) = \hat{U}(t)\psi(x,0) = e^{-i\hat{H}t/\hbar}\psi(x,0)$$
$$= e^{-i[\hat{p}_x^2/2m + V(x) - e|\mathcal{E}|x]t/\hbar}\psi(x,0). \tag{5.87}$$

[20] See Equation (2.81).

Replacing x with $x + a$ in the above equation, and using $V(x + a) = V(x)$ (since the potential is periodic) and $\psi(x + a, 0) = \psi(x, 0)e^{i k(t=0) a}$ (Bloch's theorem), we obtain

$$\psi(x + a, t) = e^{-i[\hat{p}_x^2/2m + \hat{V}(x+a) - e|\mathcal{E}|(x+a)]t/\hbar} \psi(x + a, 0) \tag{5.88}$$

$$= \underbrace{e^{-i[\hat{p}_x^2/2m + \hat{V}(x) - e|\mathcal{E}|x]t/\hbar} \psi(x, 0)}_{\psi(x,t)} \, e^{i k(t=0) a} e^{i e|\mathcal{E}|at/\hbar}. \tag{5.89}$$

We therefore get a Bloch-like relation:

$$\psi(x + a, t) = \psi(x, t)e^{i k(t) a}, \tag{5.90}$$

where

$$k(t) = \frac{e|\mathcal{E}|t}{\hbar} + k(t = 0). \tag{5.91}$$

Taking the time derivative of Equation (5.91),

$$\boxed{\frac{d}{dt}[\hbar k(t)] = e|\mathcal{E}| = F}. \tag{5.92}$$

This result is remarkably simple and beautiful: it says that the effect of an external force $F = e|\mathcal{E}|$ on the electron is to change the quantity $\hbar k(t)$, which is called the crystal momentum.[21] Note that k here is the Bloch wavevector. Electrons accelerate according to the rate of change of crystal momentum ($\hbar k$) in the periodic potential. Hence, this can be thought of as the physical meaning behind the Bloch wavevector.

In further analogy to Newton's classical mechanics, we can write down the acceleration for a Bloch state ψ_{nk} by taking the time derivative of the group velocity of that state:

$$a_{nk} = \frac{d}{dt}v_{nk} = \frac{1}{\hbar}\frac{d}{dt}\left(\frac{\partial E_n}{\partial k}\right) = \frac{1}{\hbar}\frac{\partial^2 E_n}{\partial k^2}\frac{dk}{dt}. \tag{5.93}$$

Now using the equation of motion for Bloch electrons [Equation (5.92)], we can modify this expression to

$$a_{nk} = \frac{1}{\hbar^2}\frac{\partial^2 E_n}{\partial k^2} F. \tag{5.94}$$

We can write this equation in the same form as Newton's equation of motion ($F = ma$) if we introduce the effective mass of the electron as

$$m_n^*(k) := \hbar^2 \left(\frac{\partial^2 E_n}{\partial k^2}\right)^{-1}. \tag{5.95}$$

Note that the effective mass is in general a function of the Bloch index (wavevector) k. However, if the band dispersion $E_n(k)$ can be

[21] In other words, a Bloch electron in an external force F obeys Newton's equation of motion with the real momentum replaced by the crystal momentum.

approximated as a parabola $E(k) = C_0 k^2$, where C_0 is a constant, as is often the case for band edges in semiconductors, then the effective mass is a constant: $m^* = \hbar^2/2C_0$. The values of the *band-edge* effective masses $m^*(0)$ of electrons in representative semiconductors are listed in Table 7.1, together with their band gap values.

Remark 5.1 Bloch oscillation

Bloch electrons can sometimes exhibit nonintuitive behavior that is not expected for electrons in free space. One example is the phenomenon of Bloch oscillation, which occurs in the presence of a strong DC electric field. Suppose that an electron, initially at rest at the bottom ($k = 0$) of a 1D conduction band $E_n(k)$, is accelerated by a uniform DC electric field $\mathcal{E}$. According to Equation (5.91), which is the solution to the equation of motion for Bloch electrons [Equation (5.92)], the Bloch wavenumber k increases linearly with t, moving away from the center of the FBZ ($k = 0$), i.e., $k(t) = e|\mathcal{E}|t/\hbar$. In other words, k seems to keep increasing indefinitely toward infinity!

Figure 5.12 A Bloch electron in a DC electric field oscillates both in k-space and real space, emitting electromagnetic radiation with frequency ω_B. Note that the $[\pi/a, 2\pi/a]$ portion of the band dispersion is equivalent to the $[-\pi/a, 0]$ portion, and thus the electron accelerated to the FBZ edge *disappears* at $k = \pi/a$ and immediately *reappears* at $k = -\pi/a$. © Deyin Kong (Rice University).

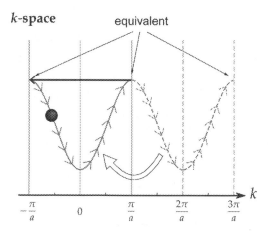

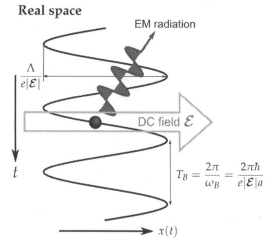

Recall, however, that all information about Bloch electron states is carried by k values within the FBZ, $k \in [-\pi/a, \pi/a]$, where a is the lattice constant. Since $e^{i(k+2\pi n/a)a} = e^{ika}$, where n is an integer, k and $k' = k + 2\pi n/a$ are equivalent.[22] Most relevantly, we see that the two FBZ edges $k = -\pi/a$ and $k' = \pi/a$ are equivalent. Therefore, when the electron reaches the right-hand edge of the FBZ, $k' = \pi/a$, it will "disappear" there and then reemerge at the left-hand edge, $k = -\pi/a$.[23] After reemerging, the electron's k value will continue to change linearly with t owing to the electric field, but this time it will move *toward* the zone center ($k = 0$). In the absence of scattering, it will eventually reach the zone center. Namely, the electron motion is periodic!

[22] See Section 5.4.1.

[23] Alternatively, we can say that the electron is Bragg scattered from the π/a state to the $-\pi/a$ state. See Section 7.2.

Let us try to be more quantitative. The electron moves in k-space at a constant "velocity" of $e|\mathcal{E}|/\hbar$, and the "distance" between the two edges of the FBZ is $2\pi/a$. Therefore, the time it takes for the electron to undergo one cycle of Bloch oscillation is

$$T_B = \frac{2\pi}{a} \times \frac{\hbar}{e|\mathcal{E}|} = \frac{2\pi}{\omega_B}, \tag{5.96}$$

where

$$\omega_B := \frac{e|\mathcal{E}|a}{\hbar} \tag{5.97}$$

is known as the Bloch frequency. By associating the group velocity of the Bloch electron, $v = \hbar^{-1}dE/dk$, with $\dot{x} = dx/dt$ and noting that $dk/dt = e|\mathcal{E}|/\hbar$, we obtain

$$x(t) = x(0) + \frac{1}{\hbar}\int_0^t \frac{dE(k(t'))}{dk}dt' = x(0) + \frac{1}{e|\mathcal{E}|}\int_0^{k(t)} \frac{dE(k')}{dk'}dk'$$

$$= x(0) + \frac{1}{e|\mathcal{E}|}\{E(k(t)) - E(0)\}, \tag{5.98}$$

where we have assumed that $k(0) = 0$.

To gain more insight into the motion of the electron in real space, let us use a specific band dispersion, $E(k) = \Lambda[1 - \cos(ka)]$, where Λ is a positive constant specifying the bandwidth (the energy difference between the bottom and the top of the band).[24] Then, Equation (5.98) leads to

[24] See the tight-binding model to be described in Section 7.1.

$$x(t) - x(0) = \frac{\Lambda}{e|\mathcal{E}|}[1 - \cos(\omega_B t)]. \tag{5.99}$$

This equation tells us that the electron position is oscillating at the Bloch frequency, and therefore that the electron is localized in space, despite the fact that the applied electric field $\mathcal{E}$ is constant in time! The localization length (i.e., the oscillation amplitude), $\Lambda/e|\mathcal{E}|$, decreases as the field strength $|\mathcal{E}|$ increases. Furthermore, since the electron is charged, its oscillatory motion should lead to the generation of electromagnetic radiation at the Bloch frequency. See Figure 5.12.

In reality, the Bloch oscillation is difficult to observe,[25] primarily because electrons in real solids are subject to frequent scattering. They are scattered by the vibrations of the atoms of the crystal (which are sometimes called phonons), by imperfections (such as defects and impurities),

[25] See also Exercise 5.13.

and by other electrons. The rate at which electrons are scattered in typical metals can be estimated to be $> 10^{12}$ Hz, from their electrical conductivity values. On the other hand, if we take the lattice constant of copper (0.36 nm), for example, and apply an electric field of 10 kV cm^{-1} (which is a strong field), the Bloch frequency is calculated to be only $\sim 10^{11}$ Hz. Therefore, in this case electrons are scattered (and therefore cease the coherent motion induced by the electric field) before completing a Bloch oscillation. This problem can be circumvented by using artificial semiconductor quantum structures called *superlattices*.[26] In a superlattice, the lattice constant a can be engineered to be large (e.g., > 10 nm), which will naturally increase the Bloch frequency $\omega_B = e|\mathcal{E}|a/\hbar$ for a given $|\mathcal{E}|$. Indeed, several groups have observed Bloch oscillations in semiconductor superlattices.[27]

[26] L. Esaki and R. Tsu, *IBM Journal of Research and Development* **14**, 61 (1970). See also Section 7.4.

[27] J. Feldmann *et al.*, *Physical Review B* **46**, 7252 (1992); K. Leo *et al.*, *Solid State Communications* **84**, 943 (1992); and C. Waschke *et al.*, *Physical Review Letters* **70**, 3319 (1993).

5.6   Chapter Summary

In this chapter, we learned the basic methods of modeling electronic states and dynamics in a given potential landscape, i.e., an engineered potential energy as a function of position, $V = V(r)$. Particular attention was given to unbound states, whose amplitudes do not vanish at infinity. Unlike the case of bound states (which we studied extensively in Chapters 2–4), we realized the fundamental importance of the probability current density $j(r)$ [Equation (5.9)], instead of the probability density $|\psi(r)|^2$, for the case of unbound states. We were able to calculate the transmission and reflection probabilities for electron waves incident onto various potential barriers, including quantum tunneling barrier potentials, while conserving the probability current density. We introduced the propagation matrix method (Section 5.3.2), which is useful when one deals with complicated structures with multiple interfaces, as are often encountered in modern semiconductor optoelectronic devices. We applied the propagation matrix method to spatially periodic potentials and found the central theorem of solid state physics, the Bloch theorem (Section 5.4.1). The theorem leads naturally to the concepts of bands, band gaps, Bloch wavevectors, and effective mass. In deriving the equation of motion for Bloch electrons in an external electric field, we were able to provide a physical meaning for the Bloch index while explaining the exotic phenomenon of Bloch oscillations.

5.7   Exercises

Exercise 5.1 (Probability current density)

(a) Show that $j(x,t) = 0$ for a bound state such as $\psi(x) = Ae^{-\alpha x}$, where A is a normalization constant and α is a positive constant.

(b) The probability current density $j(x,t)$ can be nonzero for a *superposition* of bound states. Calculate the probability current for $\Psi(x,t) = \frac{1}{\sqrt{2}}[\Psi_1(x,t) + \Psi_2(x,t)]$, where

$$\Psi_n(x,t) = \sqrt{\frac{2}{L}} \sin\left(\frac{n\pi x}{L}\right)e^{-i\omega_n t} \qquad (5.100)$$

is the time-dependent nth eigenstate of a square quantum well with width L and energy $E_n = \hbar\omega_n$, shown in Figure 2.4.[28]

[28] See also Remark 2.1.

Exercise 5.2 (Probability current density)

Consider the dynamics of a particle of mass m that moves freely inside a potential well defined by Equation (2.38). The particle is initially in the state

$$\Psi(x,0) = \sqrt{\frac{6}{5L}} \sin\left(\frac{3\pi x}{L}\right) + \sqrt{\frac{4}{5L}} \sin\left(\frac{5\pi x}{L}\right). \qquad (5.101)$$

(a) Express $\Psi(x,0)$ as a superposition of eigenfunctions $\psi_n(x)$, Equation (2.43).

(b) Find the wavefunction at any later time t, $\Psi(x,t)$.

(c) Calculate the probability density $P(x,t) = |\Psi(x,t)|^2$ and the probability current density $j(x,t)$.

(d) Verify that the probability is conserved, i.e., Equation (5.7) is satisfied.

Exercise 5.3 (Transmission probability)

(a) Express the transmission probability P_t for the step-function potential, given by Equation (5.14), in terms of the normalized incident energy $\varepsilon := E/(V_2 - V_1)$. Plot P_t versus ε for $V_1 = 0$ and $E > V_2$.

(b) When the effective masses in Regions 1 and 2 in Figure 5.1 satisfy $m_1 = 5m_2$, plot the transmission probability P_t as a function of normalized incident energy $\varepsilon := E/(V_2 - V_1)$ for $V_1 = 0$ and $E > V_2$.

Exercise 5.4 (Transmission resonances)

A beam of particles with mass m and energy E (> 0) is incident on a potential well given by

$$V(x) = \begin{cases} 0, & |x| \geq L/2, \\ -V_0, & |x| < L/2. \end{cases} \qquad (5.102)$$

(a) Derive expressions for the reflection probability (P_r) and transmission probability (P_t).

(b) Obtain the condition for vanishing reflectivity (i.e., $P_r = 0$ and $P_t = 1$).

(c) Plot P_r and P_t as a function of E from 0 to 0.5 eV, when $m = 9.11 \times 10^{-31}$ kg, $V_0 = 50$ meV, and $L = 2$ nm.

Exercise 5.5 (Quantum tunneling)

Refer to Figure 5.2 when solving the following problems.

(a) When $V_0 = 1$ eV, $L = 0.15$ nm, and $E = 0.20$ eV, estimate the tunneling probability P_t.

(b) When $V_0 = 2$ eV and $E = 0.25$ eV, plot P_t as a function of L.

(c) In a certain semiconductor device, it is desired to maintain at a low value of the leakage current associated with quantum tunneling through a rectangular potential barrier, specifically keeping $P_t < 10^{-5}$. If $\varepsilon = 0.1$, what is required for V_0 and L?

(d) Discuss how the transmission probability P_t generally depends on the effective mass of the incident electron. Then, specifically, calculate P_t when $\varepsilon = 0.1$, $V_0 = 1$ eV and $L = 1$ nm for $m^* = 0.067 m_e$ (GaAs) and $m^* = 0.014 m_e$ (InSb).

Exercise 5.6 (Effective mass ratio in tunneling)

Consider a particle with energy E (> 0) incident from the left on a square potential barrier with height V_0 and width L. The effective mass of the semiconductor material in Regions 1 ($x < 0$) and 3 ($x > L$) is m_1^*, while that in Region 2 ($0 < x < L$) is m_2^*. Calculate the transmission probability P_t as a function of energy E and discuss how $P_t(E)$ depends on the effective mass ratio m_1^*/m_2^*.

Exercise 5.7 (Scanning tunneling microscopy (STM))

The principle of STM is shown schematically in Figure 5.13(a). A single-atom tip is placed close to a conductive sample surface. When

a small bias is applied, there will be a tunneling current that is extremely sensitive to the separation between the tip and the surface.[29]
By adjusting the tip height and controlling the tunneling current to
be constant, one can map out the spatial height fluctuations of the
sample surface with atomic resolution.

A simplified 1D model of STM is shown in Figure 5.13(b). Suppose that the electron potential energies of the tip and sample are the
same, the incident electron energy is E, the vacuum level (the height
of the potential barrier) is V_0, and the vacuum gap separation is L.

[29] See Example 5.1.

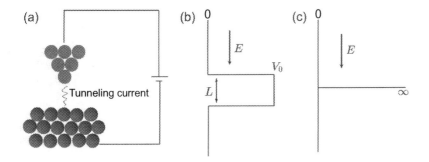

Figure 5.13 A schematic diagram of
an STM setup (a). The vacuum gap
between the metal tip and the material surface is modeled as a square
potential barrier in (b) and as a δ-
function barrier in (c).

(a) For $\varepsilon = E/V_0 = 0.1$, plot the transmission probability, P_t, as a
function of $\eta = \sqrt{2mV_0}L/\hbar$, with η in the range $1 < \eta < 5$. From
this result, state the reason why the tunneling current in an STM
is sensitive to the gap length, L.

(b) As shown in Figure 5.13(c), when the vacuum gap is small and
the vacuum energy level is high, we can approximately treat the
square potential barrier as a δ-function potential barrier $V = U_0\delta(x)$. Solve the Schrödinger equation with this potential and
calculate P_t for an electron. (*Hint*: Integrate both sides of the
Schrödinger equation.)

(c) Show analytically that when $V_0 \to \infty$, $L \to 0$, and $U_0 = V_0L$,
the delta-function potential V can account well for the property
of the square potential in terms of electron transmission. Choose
specific values and plot P_t for both potentials to show that they
are close in the above limit.

Exercise 5.8 (1D scattering potentials)

(a) Find the width L and height V_1 needed to suppress the reflection
of an electron wave of energy E ($> V_2$) in Figure 5.14. This is a
quantum mechanical "anti-reflection coating."

Figure 5.14 Potential energy engineering for propagating electrons, in analogy to the anti-reflection coating of a material surface for an incident optical wave. © Deyin Kong (Rice University).

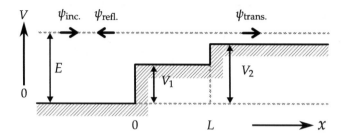

(b) Heterojunction bipolar transistors are semiconductor devices with potential energy profiles as shown in Figure 5.15. Electrons with mass m^* and energy E are injected into the device from the emitter ($x < 0$) to the base ($0 \leq x \leq L$) and propagates through the collector ($x > L$). Use the propagation matrix method to calculate the probability for transmission $P_t(E)$. What is the value of P_t when $E = 9V_0$, $V_0 = 0.2$ eV, $m^* = 0.067m_e$, $m_e = 9.11 \times 10^{-31}$ kg, and $L = 10$ nm?

Figure 5.15 Potential energy engineering for a heterojunction bipolar transistor. © Deyin Kong (Rice University).

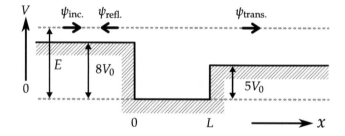

Exercise 5.9 (Josephson tunneling)

A Josephson junction (JJ) consists of two superconductors coupled by a weak link. The weak link is a thin insulating barrier, so a JJ is also known as a superconductor–insulator–superconductor (S-I-S) junction; see Figure 5.16. The Josephson effect refers to the macroscopic quantum tunneling that occurs in JJs.[30]

Superconductor 1 (2) has a macroscopic wavefunction of Cooper pairs[31] $\psi_1 = \sqrt{n_1}\, e^{i\phi_1}$ ($\psi_2 = \sqrt{n_2}\, e^{i\phi_2}$). In the presence of an electric potential difference across the junction, V, the Schrödinger equation for this two-state quantum system can be written as

$$i\hbar \frac{\partial}{\partial t} \begin{bmatrix} \sqrt{n_1}\, e^{i\phi_1} \\ \sqrt{n_2}\, e^{i\phi_2} \end{bmatrix} = \begin{bmatrix} eV & \zeta \\ \zeta & -eV \end{bmatrix} \begin{bmatrix} \sqrt{n_1}\, e^{i\phi_1} \\ \sqrt{n_2}\, e^{i\phi_2} \end{bmatrix} \tag{5.103}$$

where the constant ζ is the coupling strength of the two superconductors, which depends on the characteristics of the junction.

[30] Predicted in 1962 by the British physicist Brian David Josephson, who received the Nobel Prize in Physics in 1973. This effect is a manifestation of tunneling by superconducting Cooper pairs and is used in applications such as superconducting quantum interference devices (SQUIDs) and superconducting qubits.

[31] A Cooper pair is a pair of electrons bound together through attractive interaction mediated by lattice phonons, introduced by Leon Cooper in explaining superconductivity; L. N. Cooper, "Bound electron pairs in a degenerate Fermi gas," *Physical Review* **104**, 1189 (1956).

(a) By noting that

$$\frac{\partial}{\partial t}\left(\sqrt{n_v}\,e^{i\phi_v}\right) = \left(\frac{\partial \sqrt{n_v}}{\partial t} + i\sqrt{n_v}\,\frac{\partial \phi_v}{\partial t}\right)e^{i\phi_v}, \qquad (5.104)$$

where $v = 1, 2$, derive the following:

$$\frac{\partial \sqrt{n_1}}{\partial t} + i\sqrt{n_1}\,\frac{\partial \phi_1}{\partial t} = \frac{1}{i\hbar}\left(eV\sqrt{n_1} + \zeta\sqrt{n_2}\,e^{i\Delta\phi}\right), \qquad (5.105)$$

and its complex conjugate

$$\frac{\partial \sqrt{n_1}}{\partial t} - i\sqrt{n_1}\,\frac{\partial \phi_1}{\partial t} = \frac{-1}{i\hbar}\left(eV\sqrt{n_1} + \zeta\sqrt{n_2}\,e^{-i\Delta\phi}\right), \qquad (5.106)$$

where $\Delta\phi := \phi_2 - \phi_1$ is the phase difference across the junction, known as the Josephson phase.

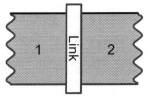

Figure 5.16 A Josephson junction. "1" and "2" represent superconductors, while the narrow region in the middle is the weak link between the two superconducting regions. © Deyin Kong (Rice University).

(b) By combining Equations (5.105) and (5.106), show the following:

$$\frac{\partial n_1}{\partial t} = -\frac{\partial n_2}{\partial t} = \frac{2\zeta\sqrt{n_1 n_2}}{\hbar}\sin\Delta\phi, \qquad (5.107)$$

$$\frac{\partial \phi_1}{\partial t} = -\frac{1}{\hbar}\left(eV + \zeta\sqrt{\frac{n_2}{n_1}}\cos\Delta\phi\right), \qquad (5.108)$$

$$\frac{\partial \phi_2}{\partial t} = \frac{1}{\hbar}\left(eV - \zeta\sqrt{\frac{n_1}{n_2}}\cos\Delta\phi\right). \qquad (5.109)$$

(c) By noting that the time derivative of the charge carrier density $\partial n_1/\partial t = -\partial n_2/\partial t$ is proportional to the current I and assuming that $n_1 \approx n_2$, derive the Josephson equations:[32]

$$I(t) = I_c \sin\Delta\phi(t), \qquad (5.110)$$

$$\frac{\partial \Delta\phi(t)}{\partial t} = \frac{2\pi}{\Phi_0}V(t), \qquad (5.111)$$

where I_c is a constant (the critical current), $\Phi_0 = h/2e$ is called the magnetic flux quantum, and $V(t)$ and $I(t)$ are the voltage across the junction and the current through the junction, respectively.

[32] When no voltage is applied ($V = 0$), the second equation shows that the phase difference is constant, which, through the first equation, means that there is a constant current! On the other hand, when a DC voltage V_{DC} is applied, the second equation shows that $\Delta\phi = 2\pi V_{DC}t/\Phi_0$, which, via the first equation, tells us that there is an AC current with a fixed frequency!

Exercise 5.10 (Propagation matrix)

Starting from the conservation of probability current density between the initial layer and the final layer, $j_1 = j_n$, derive the following property of propagation matrices: $\det \mathbf{P} = 1$.

Exercise 5.11 (Double-well potential)

Use the propagation matrix method to analyze the transmission and reflection probabilities for a flux of electrons with energy $E\,(> 0)$ coming from $x = -\infty$ and incident onto the double-well potential shown in Figure 5.17.

Figure 5.17 Double-well potential.
© Deyin Kong (Rice University).

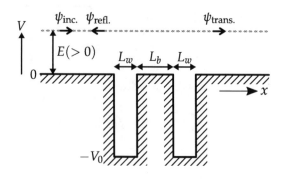

Exercise 5.12 (Effective mass and velocity)

Calculate the effective mass $m^*(k)$ and velocity $v(k)$ of a Bloch electron at k in a band whose dispersion is $E(k) = \gamma[1 - \cos(kL)]$, where γ and L are constants. Where in k-space is the energy a maximum, and where is it a minimum? What is the band-edge mass, $m^*(0)$?

Exercise 5.13 (Bloch electron dynamics)

A semiconductor has an $E(k)$ relationship in the conduction band given by

$$E(k) = 1.0 + 1.0 \times 10^{-15}k^2 - 1.267 \times 10^{-31}k^4, \qquad (5.112)$$

where k is given in cm^{-1}, the first constant (1.0×10^{-15}) is in eV cm^2, the second constant is in eV cm^4, and E is in eV.

(a) Plot this relationship for $|k| \leq 2\pi \times 10^7$ cm^{-1}.

(b) What is the effective mass (in units of $m_e = 9.11 \times 10^{-31}$ kg) for electrons at the bottom of the conduction band?

(c) An electric field of strength 10^3 V cm^{-1} is applied to the sample. How rapidly will k increase with time $(\partial k/\partial t)$? The answer should have a unit of cm^{-1} s^{-1}.

(d) Assuming that $|k| \leq 2\pi \times 10^7$ cm^{-1} defines the first Brillouin zone, how long will it take for the electron to go from the band bottom to the Brillouin zone edge?

(e) The electrons in this sample have a mean scattering time $\tau_s = 10^{-12}$ s. What is the mean group velocity $\langle v_g \rangle$ for the electron wave packets? Note: with such a short scattering time, the electrons will be forced to stay near the bottom of the band, where k is small, and hence the k^4 term can probably be ignored compared with the k^2 term, in Equation (5.112).

(f) What is the mobility, μ_e,[33] of the electrons near the bottom of the band?

(g) What average velocity do you get using $\langle v \rangle = \mu_e \mathcal{E}$ for an electric field of 10^3 V cm^{-1}? How does this compare with the mean group velocity for electron wavepackets obtained in part (c)?

6

Light–Matter Interaction

Learning objectives:

- Becoming familiar with the time-dependent Schrödinger equation.

- Learning the basic ideas of perturbation theory.

- Calculating transition probabilities using Fermi's golden rule.

- Deriving the optical Bloch equations using the density matrix formalism.

- Understanding the concepts of population inversion, gain, and lasing, using rate equations.

THE THEORY OF THE INTERACTION OF RADIATION WITH MATTER is fundamentally important for describing how modern semiconductor devices generate, detect, and modulate light. These devices, known as optoelectronic devices, are behind today's technology in diverse areas, including communications, imaging, spectroscopy, sensing, and energy harvesting. They may also become essential components in future quantum technology based on photons. In this chapter we will learn the basic theoretical formalism for describing light–matter interaction phenomena, starting from microscopic processes such as absorption, spontaneous emission, and stimulated emission and ending with the conditions for achieving gain, which is a fundamental requirement for a laser.

6.1 Time Evolution of a Driven Two-Level System

The overall goal of this chapter is to understand the basic quantum mechanical formalism of light–matter interaction. The time-dependent Schrödinger equation[1] we need to solve is

$$i\hbar\frac{\partial}{\partial t}|\Psi(\bm{r},t)\rangle = \hat{H}(t)|\Psi(\bm{r},t)\rangle, \tag{6.1}$$

where $\hat{H}(t) = \hat{H}_0 + \hat{W}(t)$, $\hat{H}_0 = \hat{\bm{p}}^2/2m + V(\bm{r})$ is the unperturbed matter Hamiltonian, and $\hat{W}(t)$ represents the light–matter interaction.

Let us assume that $\hat{H}_0$ has a complete orthonormal set of eigenstates, $|n\rangle$,[2] with corresponding eigenenergies, E_n, such that

$$\hat{H}_0|n\rangle = E_n|n\rangle, \tag{6.2}$$

$$\langle m|n\rangle = \delta_{mn}, \tag{6.3}$$

$$\sum_n |n\rangle\langle n| = \hat{1}. \tag{6.4}$$

[1] Equation (2.79).

[2] This $|n\rangle$ is *not* the number state described in Remark 3.2. Rather, it is simply the nth eigenstate of the time-independent Hamiltonian $\hat{H}_0$.

The total wavefunction for the nth eigenstate is then given by

$$|\Psi_n\rangle = |n\rangle e^{-i\omega_n t}, \tag{6.5}$$

where $\omega_n := E_n/\hbar$. We can expand $|\Psi\rangle$ as a linear combination of $|\Psi_n\rangle$, i.e.,[3]

$$|\Psi\rangle = \sum_n c_n(t)|\Psi_n\rangle. \tag{6.6}$$

By substituting Equation (6.6) into Equation (6.1) and multiplying both sides by $\langle m|$, we derive

$$i\hbar\dot{c}_m = \sum_n c_n W_{mn} e^{i\omega_{nm}t}, \tag{6.7}$$

where $W_{mn} := \langle m|\hat{W}|n\rangle$ and $\omega_{mn} := \omega_m - \omega_n$. We have not made any approximations, and Equation (6.7) is fully equivalent to the original time-dependent Schrödinger equation [Equation (6.1)].

To understand the time evolution of the system described by Equation (6.7), let us apply it to a physical situation where a two-level approximation is valid. Namely, the eigenstates of the system are $|\Psi_1\rangle = |1\rangle e^{-i\omega_1 t}$ and $|\Psi_2\rangle = |2\rangle e^{-i\omega_2 t}$ with eigenenergies $E_1 = \hbar\omega_1$ and $E_2 = \hbar\omega_2 (> E_1)$, respectively. Let us also define $\omega_0 := \omega_2 - \omega_1 = \omega_{21} = -\omega_{12}$. Furthermore, we assume an electric dipole interaction,[4] for which $W_{11} = W_{22} = 0$; we also have $W_{21} = W_{12}^*$.

With these assumptions and definitions, Equation (6.7) for this two-level system simplifies to

$$i\hbar\dot{c}_1 = c_2 W_{12} e^{-i\omega_0 t} \quad \text{and} \quad i\hbar\dot{c}_2 = c_1 W_{12}^* e^{i\omega_0 t}. \tag{6.8}$$

We seek solutions in the following form:

$$c_1 = b_1 \exp\left[i\left(\omega - \frac{\omega_0}{2}\right)t\right] \quad \text{and} \quad c_2 = b_2 \exp\left[i\left(\omega + \frac{\omega_0}{2}\right)t\right]. \tag{6.9}$$

Upon substitution of Equations (6.9) into Equations (6.8), we find the following coupled equations to be solved for b_1 and b_2:

$$\begin{bmatrix} \hbar\left(\omega - \dfrac{\omega_0}{2}\right) & W_{12} \\ W_{12}^* & \hbar\left(\omega + \dfrac{\omega_0}{2}\right) \end{bmatrix} \begin{bmatrix} b_1 \\ b_2 \end{bmatrix} = 0. \tag{6.10}$$

For nontrivial solutions to exist for b_1 and b_2, the determinant of the coefficient matrix has to vanish, i.e.,

$$\begin{vmatrix} \hbar\left(\omega - \dfrac{\omega_0}{2}\right) & W_{12} \\ W_{12}^* & \hbar\left(\omega + \dfrac{\omega_0}{2}\right) \end{vmatrix}^2 = \hbar^2\left(\omega - \frac{\omega_0}{4}\right) - |W_{12}|^2 = 0, \tag{6.11}$$

[3] Note that the expansion coefficients are now time dependent. We can interpret $|c_n(t)|^2$ as the probability that the system is found to be in the state $|\Psi_n\rangle$ at time t.

[4] That is, $\hat{W} \propto x$, so $W_{nn} \propto \langle n|x|n\rangle = 0$ and $W_{nm} \propto \langle n|x|m\rangle = \langle xn|m\rangle = \langle m|xn\rangle^* = \langle m|x|n\rangle^*$. See Section 6.2.2 for more details about electric dipole interactions.

which yields

$$\omega = \pm\Omega, \tag{6.12}$$

$$\Omega := \sqrt{\frac{|W_{12}|^2}{\hbar^2} + \left(\frac{\omega_0}{2}\right)^2}. \tag{6.13}$$

The frequency Ω is known as the generalized Rabi frequency. There are two specific solutions for c_1 and c_2, corresponding to the cases $\omega = \Omega$ and $\omega = -\Omega$, and thus the general solution is a linear combination of the two, i.e.,

$$c_1 = (b_1^+ e^{i\Omega t} + b_1^- e^{-i\Omega t})e^{-i\omega_0 t/2}, \tag{6.14}$$

$$c_2 = (b_2^+ e^{i\Omega t} + b_2^- e^{-i\Omega t})e^{i\omega_0 t/2}. \tag{6.15}$$

The coefficients b_1^+, b_1^-, b_2^+, and b_2^- are to be determined by the initial conditions.

Let us consider the following initial conditions: $c_1(0) = 1$ and $c_2(0) = 0$.[5] Using these initial conditions in Equations (6.14), (6.15), and (6.10), we obtain

$$b_1^+ + b_1^- = 1, \tag{6.16}$$

$$b_2^+ + b_2^- = 0, \tag{6.17}$$

$$\hbar\left(\Omega - \frac{\omega_0}{2}\right)b_1^+ + W_{12}b_2^+ = 0, \tag{6.18}$$

$$\hbar\left(-\Omega - \frac{\omega_0}{2}\right)b_1^- + W_{12}b_2^- = 0. \tag{6.19}$$

By solving these four coupled equations, we get

$$b_1^\pm = \frac{1}{2} \pm \frac{\omega_0}{4\Omega}, \tag{6.20}$$

$$b_2^\pm = \mp\frac{W_{21}}{2\hbar\Omega}. \tag{6.21}$$

Hence, we can calculate $c_1(t)$ and $c_2(t)$ to be

$$c_1 = \left[\left(\frac{1}{2} + \frac{\omega_0}{4\Omega}\right)e^{i\Omega t} + \left(\frac{1}{2} - \frac{\omega_0}{4\Omega}\right)e^{-i\Omega t}\right]e^{-i\omega_0 t/2}$$

$$= \left(\cos\Omega t + i\frac{\omega_0}{2\Omega}\sin\Omega t\right)e^{-i\omega_0 t/2}, \tag{6.22}$$

$$c_2 = \left(-\frac{W_{21}}{2\hbar\Omega}e^{i\Omega t} + \frac{W_{21}}{2\hbar\Omega}e^{-i\Omega t}\right)e^{i\omega_0 t/2}$$

$$= -i\frac{W_{21}}{\hbar\Omega}\sin\Omega t\, e^{i\omega_0 t/2}. \tag{6.23}$$

The squared coefficients $|c_1|^2$ and $|c_2|^2$ are plotted as functions of t for different values of $|W_{12}|/\hbar\omega_0$ in Figure 6.1, where the system is periodically oscillating between the two states. These oscillations are known as Rabi oscillations. We see that as $|W_{12}|$ – a measure of the light–matter coupling strength – increases, the oscillation amplitude increases.

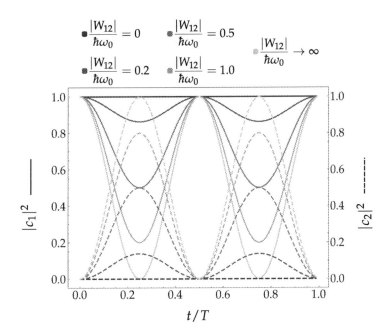

$\dfrac{|W_{12}|}{\hbar\omega_0} = 0$ $\dfrac{|W_{12}|}{\hbar\omega_0} = 0.5$

$\dfrac{|W_{12}|}{\hbar\omega_0} = 0.2$ $\dfrac{|W_{12}|}{\hbar\omega_0} = 1.0$

$\dfrac{|W_{12}|}{\hbar\omega_0} \to \infty$

Figure 6.1 Rabi oscillations in a coherently driven two-level system. The population probabilities of the two states, $|c_1|^2$ and $|c_2|^2$, are plotted as functions of t/T, where $T = 2\pi/\Omega$, for different values of $|W_{12}|/\hbar\omega_0$. © Deyin Kong (Rice University).

6.2 Transition Probability

Equation (6.7) represents coupled equations to be solved for the coefficients $c_m(t)$, and once we have the $c_m(t)$ we know the probability of finding the system in the state $|m\rangle$ at time t, $|c_m(t)|^2$. Let us use the initial conditions

$$c_m(0) = \delta_{mi}, \qquad (6.24)$$

that is, the initial state of the system is given by $|\Psi(0)\rangle = |i\rangle$. What we then wish to calculate is

$$P_{i \to f}(t) = |c_f(t)|^2, \qquad (6.25)$$

which can be interpreted as the transition probability from the initial state $|i\rangle$ to the final state $|f\rangle$.

6.2.1 Time-Dependent Perturbation Theory

Here we will use a perturbation method[6] to calculate $P_{i \to f}(t)$. We replace W_{mn} with $\alpha_0 W_{mn}$ in Equation (6.7) and use the small dimensionless parameter α_0 ($\ll 1$) to expand c_m as

$$c_m(t) = c_m^{(0)}(t) + \alpha_0 c_m^{(1)}(t) + \alpha_0^2 c_m^{(2)}(t) + \cdots \qquad (6.26)$$

[6] Namely, we assume that the light–matter interaction only slightly modifies the matter state and energy.

Substituting Equation (6.26) into Equation (6.7), we get

$$i\hbar\left[\dot{c}_m^{(0)} + \alpha_0\dot{c}_m^{(1)} + \alpha_0^2\dot{c}_m^{(2)} + \cdots\right]$$
$$= \sum_n\left[c_n^{(0)} + \alpha_0 c_n^{(1)} + \alpha_0^2 c_n^{(2)} + \cdots\right]\alpha_0 W_{mn}e^{i\omega_{mn}t}. \tag{6.27}$$

[7] This simply means that the zeroth-order term (i.e., the unperturbed term) $c_m^{(0)}$ is given by the initial value, i.e., $c_m^{(0)} = \delta_{mi}$.

First, by equating the zeroth-order terms on the left- and right-hand sides of this equation, we find[7]

$$i\hbar\dot{c}_m^{(0)} = 0 \quad\rightarrow\quad c_m^{(0)}(t) = \text{constant}. \tag{6.28}$$

Next, by comparing the first-order terms on the left- and right-hand sides, we obtain

$$i\hbar\dot{c}_m^{(1)} = \sum_n c_n^{(0)} W_{mn}e^{i\omega_{mn}t} = \sum_n \delta_{ni}W_{mn}e^{i\omega_{mn}t} = W_{mi}e^{i\omega_{mi}t}. \tag{6.29}$$

We can integrate this equation to obtain

$$c_m^{(1)}(t) = \frac{1}{i\hbar}\int_0^t W_{mi}(t')e^{i\omega_{mi}t'}dt'. \tag{6.30}$$

Therefore, the transition probability from $|i\rangle$ to $|f\rangle$ at time t calculated within first-order perturbation theory is

$$P_{i\to f}(t) = |c_f^{(1)}(t)|^2 = \frac{1}{\hbar^2}\left|\int_0^t W_{fi}(t')e^{i\omega_{fi}t'}dt'\right|^2. \tag{6.31}$$

Let us specifically consider a perturbation Hamiltonian due to a light wave with frequency ω;

$$\hat{W} = \hat{W}'(e^{i\omega t} + e^{-i\omega t}) = 2\hat{W}'\cos\omega t, \tag{6.32}$$

[8] See Section 6.2.2.

where the exact form of $\hat{W}'$ will be determined later.[8] Correspondingly, the transition probability from $|i\rangle$ to $|f\rangle$ at time t is written

$$P_{i\to f}(t) = \frac{|W'_{fi}|^2}{\hbar^2}\left|\int_0^t[e^{i(\omega_{fi}+\omega)t'} + e^{i(\omega_{fi}-\omega)t'}]dt'\right|^2$$
$$= \frac{|W'_{fi}|^2}{\hbar^2}\left|\frac{e^{i(\omega_{fi}+\omega)t} - 1}{i(\omega_{fi}+\omega)} + \frac{e^{i(\omega_{fi}-\omega)t} - 1}{i(\omega_{fi}-\omega)}\right|^2. \tag{6.33}$$

[9] The "sinc" function is defined as $\text{sinc}(x) = \sin x/x$.

For near-resonance excitation ($\omega \sim \omega_{fi}$), the first term inside the verticals can be neglected. Hence,[9]

$$P_{i\to f}(t) = \frac{|W'_{fi}|^2}{\hbar^2}\left|\frac{e^{i(\omega_{fi}-\omega)t} - 1}{i(\omega_{fi}-\omega)}\right|^2 = \frac{|W'_{fi}|^2}{\hbar^2}\frac{2[1 - \cos\{(\omega_{fi}-\omega)t\}]}{(\omega_{fi}-\omega)^2}$$
$$= \frac{|W'_{fi}|^2}{4\hbar^2}t^2\frac{\sin^2\left\{\frac{(\omega_{fi}-\omega)t}{2}\right\}}{\left\{\frac{(\omega_{fi}-\omega)t}{2}\right\}^2} = \frac{|W'_{fi}|^2}{4\hbar^2}t^2\text{sinc}^2\left\{\frac{(\omega_{fi}-\omega)t}{2}\right\}. \tag{6.34}$$

6.2.2   Dipole Matrix Element

The first factor appearing in the expression that we obtained for the transition probability $P_{i \to f}$ [Equation (6.34)] is $|W'_{fi}|^2$. To find a more concrete expression for this matrix element, let us start with the classical mechanics of a charged particle interacting with an electromagnetic field within the so-called minimal-coupling approximation.[10] The minimal-coupling Lagrangian, $\mathcal{L}_{min}$, for a particle of mass m and charge q in an electromagnetic field is then given by

$$\mathcal{L}_{min} = \sum_{i=1}^{3} \frac{1}{2} m \dot{x}_i^2 + \sum_{i=1}^{3} q \dot{x}_i A_i - q\varphi, \qquad (6.35)$$

where $\dot{r} = (\dot{x}_1, \dot{x}_2, \dot{x}_3)$ is the velocity of the particle, A is the vector potential, and φ is the scalar potential.[11] The Euler–Lagrange equations of motion[12] based on this Lagrangian produce the correct Newton's equation of motion for a charged particle under the Lorentz force: $m\ddot{r} = q\mathcal{E} + q\dot{r} \times B$. The components of the canonical momentum p can be calculated from $\mathcal{L}_{min}$ as

$$p_i = \frac{\partial \mathcal{L}_{min}}{\partial \dot{x}_i} = m\dot{x}_i + qA_i, \qquad (6.36)$$

while the kinetic momentum is given by $P_i = m\dot{x}_i = p_i - qA_i$ and the Hamiltonian can be derived as

$$\mathcal{H}_{min} = \sum_{i=1}^{3} \dot{x}_i p_i - \mathcal{L}_{min} = \frac{|P|^2}{2m} + q\varphi. \qquad (6.37)$$

For an electron (with charge $-e$) in the presence of a static potential $V(r)$ and a harmonically oscillating electromagnetic wave, we have $\mathcal{H} = |p + eA|^2/2m + V(r)$ and

$$A = \frac{1}{\omega} \mathcal{E}_0 \cos \omega t = \frac{1}{2\omega} \mathcal{E}_0 (e^{i\omega t} + e^{-i\omega t}), \qquad (6.38)$$

$$\mathcal{E} = -\frac{\partial}{\partial t} A = \mathcal{E}_0 \sin \omega t, \qquad (6.39)$$

where $\mathcal{E}_0$ is a constant vector specifying the polarization direction and amplitude of the electric field wave.[13]

The quantum mechanical Hamiltonian is thus written as

$$\begin{aligned}
\hat{H} &= \frac{|\hat{P}|^2}{2m} + V(r) = \frac{1}{2m} |\hat{p} + eA|^2 + V(r) \\
&= \frac{1}{2m} [\hat{p}^2 + e(\hat{p} \cdot A + A \cdot \hat{p}) + e^2 A^2] + V(r) \\
&= \hat{H}_0 + \frac{e}{m} A \cdot \hat{p} + \frac{e^2 A^2}{2m},
\end{aligned} \qquad (6.40)$$

[10] In this approximation, multipole moments higher than dipole moments are neglected, and the spin of the particle is not taken into account.

[11] The electric and magnetic fields are expressed in terms of A and φ as $\mathcal{E} = -\nabla \varphi - \partial A/\partial t$ and $B = \nabla \times A$, respectively.

[12] $\partial \mathcal{L}/\partial x_i - (d/dt)(\partial \mathcal{L}/\partial \dot{x}_i) = 0$. See Appendix A.

[13] Under the gauge transformation $A \to A + \nabla f$ and $\varphi \to \varphi - \dot{f}$, both $\mathcal{E}$ and B remain the same, where f is an arbitrary scalar function of r and t. We choose the Coulomb gauge defined by $\nabla \cdot A = 0$. We can show that, in the Coulomb gauge, the operators A and $\hat{p}$ commute.

where $\hat{H}_0 = \hat{p}^2/2m + V(r)$ and $\hat{p} = -i\hbar\nabla$. The third term in the Hamiltonian is variously known as the A^2 term, the ponderomotive term, or the diamagnetic term, and is negligible compared with the second term as long as $A = |A|$ is small, which is true for situations where perturbation theory works well.[14] Therefore, for our purpose of calculating the transition probability within perturbation theory, we can associate $\hat{W}$ with the second term, i.e.,

$$\hat{W} = \frac{e}{m}A \cdot \hat{p}. \tag{6.41}$$

Assuming that the light wave is linearly polarized along the x-direction, we can write

$$A = \frac{1}{2\omega}\mathcal{E}_0(e^{i\omega t} + e^{-i\omega t})e_x, \tag{6.42}$$

and thus,

$$\hat{W} = \frac{e}{m}\frac{1}{2\omega}\mathcal{E}_0(e^{i\omega t} + e^{-i\omega t})\hat{p}_x = \frac{e\mathcal{E}_0}{2m\omega}\hat{p}_x(e^{i\omega t} + e^{-i\omega t}). \tag{6.43}$$

By comparing Equations (6.32) and (6.43), we conclude that $\hat{W}' = (e\mathcal{E}_0/2m\omega)\hat{p}_x$ and thus

$$W'_{fi} = \frac{e\mathcal{E}_0}{2m\omega}p_{fi}, \tag{6.44}$$

where $p_{fi} = \langle f|\hat{p}_x|i\rangle$ and $\hat{p}_x = -i\hbar\partial/\partial x$. Further, using Equation (3.170), we can show that

$$\begin{aligned}
p_{fi} &= \frac{im}{\hbar}\langle f|[\hat{H}_0, \hat{x}]|i\rangle = \frac{im}{\hbar}\langle f|\hat{H}_0\hat{x} - \hat{x}\hat{H}_0|i\rangle \\
&= \frac{im}{\hbar}(E_f - E_i)\langle f|\hat{x}|i\rangle = im\omega_{fi}x_{fi}. \tag{6.45}
\end{aligned}$$

By combining Equations (6.44) and (6.45), we find that

$$|W'_{fi}| = \frac{|\mathcal{E}_0|\,e\,|x_{fi}|}{2}\left|\frac{\omega_{fi}}{\omega}\right|. \tag{6.46}$$

Hence, when $\omega \sim \omega_{fi}$, $|W'_{fi}|$ is given approximately by the product of the electric field amplitude $|\mathcal{E}_0|$ and the dipole moment $e|x_{fi}|$.

As shown by Equation (6.46), the strength of the dipole transition between the states $|i\rangle$ and $|f\rangle$ of a given system is proportional to the (f, i) matrix element of $\hat{x}$, i.e., $|x_{fi}| = |\langle f|\hat{x}|i\rangle|$. This quantity can be expressed through a dimensionless parameter called the oscillator strength, F_{fi}, which is defined relative to the square of the matrix element between the $|0\rangle$ and $|1\rangle$ states of the 1D SHO [see Equation (3.35)], which is

$$(x_{01}^2)_{\text{SHO}} := |\langle 0_{\text{SHO}}|\hat{x}|1_{\text{SHO}}\rangle|^2 = \frac{\hbar}{2m\omega} = \frac{\hbar^2}{2mE}. \tag{6.47}$$

[14] The A^2 term becomes important for extremely nonlinear optical phenomena (also known as strong-field physics). At a laser-beam focus, a particle with charge q experiences a repulsive force $F_p = -\nabla U_p$. Here, $U_p = q^2A^2/4m$ is known as the ponderomotive potential energy (this refers to the potential energy change due to a change in position of the mass of the particle, from the Latin "ponder," meaning weight). See, e.g., *Atoms in Strong Fields*, edited by C. A. Nicolaides (Plenum Press, 1990).

The oscillator strength is then defined as[15]

$$F_{fi} = |x_{fi}|^2 \times \left(\frac{\hbar}{2m\omega_{fi}}\right)^{-1} = |\langle f|\hat{x}|i\rangle|^2 \frac{2m}{\hbar^2}(E_f - E_i). \quad (6.48)$$

When F_{fi} is finite and large, we say that the $i \to f$ transition is allowed or bright. Correspondingly, when $F_{fi} \approx 0$, the transition is said to be forbidden, not allowed, or dark. The rules that determine whether F_{fi} is finite or not are called selection rules.

6.2.3 Fermi's Golden Rule

Let us next analyze the second factor in the expression for the transition probability, $P_{i\to f}(t)$, Equation (6.34):

$$t^2 \text{sinc}^2\left\{\frac{(\omega_{fi} - \omega)t}{2}\right\} \quad (6.49)$$

The $\text{sinc}^2\left\{(\omega_{fi} - \omega)t/2\right\}$ part is plotted as a function of $\omega_{fi} - \omega$ in Figure 6.2(a). The finite width of the central peak may look counterintuitive because it indicates that there is a finite transition probability even when the light frequency ω is not exactly equal to the transition frequency ω_{fi}.[16] The reason for the finite width is that the time duration t of the light–matter interaction is finite, as shown in Figure 6.2(b), which, as a result of Heisenberg's uncertainty principle, introduces uncertainty in energy.[17] As $t \to \infty$, the central peakwidth approaches zero. In fact, as $t \to \infty$, $\text{sinc}^2\left\{(\omega_{fi} - \omega)t/2\right\} \to (2\pi/t)\delta(\omega_{fi} - \omega)$.

A common situation occurs when an incident light wave with frequency ω excites an electron from a given initial state $|i\rangle$ into a continuum of final states, $|f\rangle$, with a density of states $\rho(\omega_{fi})$, as shown in Figure 6.3. Assuming that t is long enough for us to replace $\text{sinc}^2\left\{(\omega_{fi} - \omega)t/2\right\}$ with $(2\pi/t)\delta(\omega_{fi} - \omega)$, we can evaluate the transition probability as

$$P_{i\to f}(t) = \int_{-\infty}^{\infty} \frac{|W'_{fi}|^2}{4\hbar^2} t^2 \frac{2\pi}{t}\delta(\omega_{fi} - \omega)\,\rho(\omega_{fi})d\omega_{fi}$$
$$= \frac{\pi}{2\hbar^2}|W'_{fi}(\omega_{fi})|^2\,\rho(\omega = \omega_{fi})\,t. \quad (6.50)$$

The corresponding transition rate is given by

$$\boxed{G_{i\to f} = \frac{d}{dt}P_{i\to f} = \frac{\pi}{2\hbar^2}|W'_{fi}(\omega_{fi})|^2\,\rho(\omega = \omega_{fi})}. \quad (6.51)$$

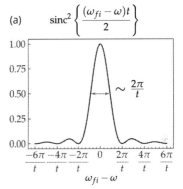

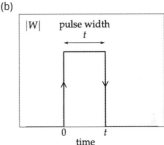

Figure 6.2 (a) $\text{sinc}^2\left\{(\omega_{fi} - \omega)t/2\right\}$ is plotted as a function of $\omega_{fi} - \omega$. The central peak width tends toward zero as $t \to \infty$. (b) Light–matter interaction strength versus time for the situation under consideration. The light is turned on at $t = 0$ and then turned off after a time t. © Deyin Kong (Rice University).

[16] This is not due to line broadening introduced by scattering or collisions, which we have so far been neglecting. Neither is it due to light-induced power broadening because the light intensity is assumed to be very small, so perturbation theory works well (see Section 6.4.4).

[17] Time t and energy E are conjugate variables satisfying $\Delta t\,\Delta E > \hbar/2$. Here, since $E = \hbar\omega$, $\Delta t\,\Delta\omega > 1/2$.

Figure 6.3 An incident light wave with frequency ω excites an electron from an initial state $|i\rangle$ into a continuum of final states with a band width of ΔE and a density of states $\rho(\omega_{fi})$. © Deyin Kong (Rice University).

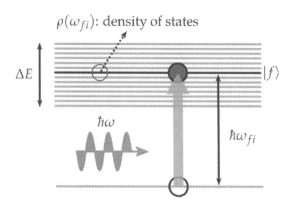

This last expression is known as Fermi's golden rule. Note that we assumed that t is long enough that the energy width $\sim 2\pi\hbar/t$ of the sinc function is much smaller than the energy bandwidth, ΔE, of the continuum of final states. Thus, $h/t \ll \Delta E$ or $t \gg h/\Delta E$ has to be satisfied. On the other hand, for perturbation theory to work, the transition probability to all final states $|f\rangle$ has to be small in comparison with unity, i.e., $|c_f(t)|^2 \ll 1$ and $|c_i(t)|^2 \sim 1$; in other words, the initial state $|i\rangle$ is not significantly depleted. From the expression for $P_{i \to f}$, this is possible when $(t/2\hbar^2)|W'_{fi}|^2 \ll 1$ or $t \ll \hbar^2/|W'_{fi}|^2$. Combining these lower and upper bounds, t has to be in the following range:

$$\frac{\hbar^2}{|W'_{fi}|^2} \gg t \gg \frac{\hbar}{\Delta E}. \tag{6.52}$$

These requirements need to be examined when one uses Fermi's golden rule in analyzing experimental results.

6.3 Density Matrix Formalism

Here we introduce a powerful method for describing the time evolution of large systems of indistinguishable particles – the density matrix formalism.[18] This is an especially convenient method for ensemble averaging, when the state of the system is not known in a precise manner, for example, when the wavefunction of each electron keeps changing through collisions. It is also straightforward to introduce relaxation times to treat collisional broadening within the density matrix formalism.

The density matrix formalism does not require precise knowledge of the wavefunction, $|\Psi(t)\rangle = \sum_n c_n(t)|n\rangle$. Namely, we do not have to have precise knowledge of $c_n(t)$ although we have to know precisely the eigenstates $|n\rangle$, which form a complete, orthonormal set.

[18] The density matrix formalism was first used in 1927 by John von Neumann and Lev Davidovich Landau independently. Von Neumann introduced it in developing the theory of quantum statistical mechanics and quantum measurements.

However, we assume that we have enough information to calculate the ensemble average[19] for c_n, i.e.,

$$\overline{|\Psi\rangle} = \sum_n \bar{c}_n(t)|n\rangle. \tag{6.53}$$

Then we can calculate the ensemble average of the expectation value of any observable A as

$$\overline{\langle A \rangle} = \overline{\langle \Psi | \hat{A} | \Psi \rangle} = \left(\sum_m \langle m | \bar{c}_m^* \right) \hat{A} \left(\sum_n \bar{c}_n(t) | n \rangle \right)$$

$$= \sum_{n,m} \langle m | \hat{A} | n \rangle \, \overline{c_m^* c_n}. \tag{6.54}$$

We define the (n, m) element of the density matrix as

$$\rho_{nm} := \overline{c_m^* c_n} = \overline{\langle \Psi | m \rangle \langle n | \Psi \rangle}$$

$$= \overline{\langle n | \Psi \rangle \langle \Psi | m \rangle} = \langle n | \hat{\rho} | m \rangle. \tag{6.55}$$

Here, $\hat{\rho}$ is the density operator.

Since $\rho_{nm} = \rho_{mn}^*$ by definition, $\hat{\rho}$ is a hermitian operator. In addition, using the density matrix, we can rewrite Equation (6.54) as

$$\boxed{\overline{\langle A \rangle} = \sum_{n,m} A_{mn} \rho_{nm} = \sum_m (A\rho)_{mm} = \mathrm{Tr}(A\rho)}. \tag{6.56}$$

This provides us with a straightforward recipe for calculating $\overline{\langle A \rangle}$ for any observable A: evaluate the trace of the product $A\rho$. Also, since the wavefunction $|\Psi\rangle$ must be normalized,

$$\mathrm{Tr}(\rho) = \sum_m \rho_{mm} = \sum_m \overline{|c_m|}^2 = 1. \tag{6.57}$$

Keep in mind that the density matrix is time dependent and can be used to describe the *time evolution of the system*. Different elements of the density matrix have distinctly different meanings, and understanding their time evolution can provide insight into different aspects of the dynamics of the system. See the box at the side. Briefly, while the on-diagonal elements provide information on the populations of the states, the off-diagonal elements are related to the *dynamics of coherence* in the system.

Using the time-dependent Schrödinger equation, the time derivative of $\hat{\rho}$ can be written as

$$\frac{d\hat{\rho}}{dt} = \frac{d\overline{|\Psi\rangle}}{dt} \overline{\langle \Psi |} + \overline{|\Psi\rangle} \frac{d\overline{\langle \Psi |}}{dt}$$

$$= \frac{1}{i\hbar} \hat{H} \overline{|\Psi\rangle \langle \Psi |} - \frac{1}{i\hbar} \overline{|\Psi\rangle \langle \Psi |} \hat{H} = \frac{1}{i\hbar} [\hat{H}, \hat{\rho}]. \tag{6.58}$$

This is the fundamental equation of motion for the density operator, $\hat{\rho}$, to be used in the following sections.

[19] Note that ensemble averages assumed here are given in a classical statistical sense, as opposed to quantum mechanical expectation values. The wavefunction $\overline{|\Psi\rangle}$ represents a system in a classical statistical ensemble of different state vectors. For example, there can be a 50% probability that $\overline{|\Psi\rangle} = |\Psi_1\rangle$ and a 50% probability that $\overline{|\Psi\rangle} = |\Psi_2\rangle$. In this sense, $\overline{|\Psi\rangle}$ is called a *mixed* state whereas $|\Psi_1\rangle$ and $|\Psi_2\rangle$ are *pure* states. Mixed states are *not* superposition states. For example, $(|\Psi_1\rangle + |\Psi_2\rangle)/\sqrt{2}$ is a superposition state, which is a pure state.

Physical meaning of density matrix elements:

- The nth on-diagonal element, $\rho_{nn} = \overline{|c_n|}^2$, provides the ensemble average of the probability that the system is in eigenstate $|n\rangle$.

- The (n, m)th off-diagonal element $(n \neq m)$, $\rho_{nm} = \overline{c_m^* c_n}$, is nonzero *only when* both $\bar{c}_n$ and $\bar{c}_m$ are nonzero, which is possible only when the system is in a superposition state of $|n\rangle$ and $|m\rangle$.

6.4   Optical Bloch Equations and Optical Susceptibility

In this section, we will use the density matrix formalism to derive the optical susceptibility,[20] $\chi(\omega) = \chi'(\omega) + i\chi''(\omega)$, of an ensemble of two-level atoms with resonance frequency $\omega_0 = \omega_{21} = (E_2 - E_1)/\hbar$ excited by an electromagnetic wave with frequency ω. Once $\chi(\omega)$ is known, we can obtain the relative permittivity (also known as the dielectric function), $\varepsilon_r(\omega)$, as

$$\varepsilon_r(\omega) = 1 + \chi(\omega), \tag{6.59}$$

[21] The quantities n_{op} and κ are the real and imaginary parts, respectively, of the complex refractive index, $\tilde{N}$. The imaginary part κ is also known as the extinction coefficient; it is related to the absorption coefficient, α, by $\kappa = \alpha c/2\omega$, where c is the speed of light. As a monochromatic light wave propagates through a distance z in an absorbing medium, its intensity decays as $e^{-\alpha z}$.

and the (complex) refractive index, $\tilde{N}(\omega)$, as[21]

$$\tilde{N}(\omega) = \sqrt{\varepsilon_r(\omega)} = \sqrt{1 + \chi(\omega)} = n_{op} - i\kappa. \tag{6.60}$$

The electric polarization, $\mathcal{P}$, of a medium is related to the applied electric field, $\mathcal{E}$, through $\mathcal{P} = \varepsilon_0 \chi \mathcal{E}$, where $\varepsilon_0 = 8.8541878128 \times 10^{-12}\,\mathrm{F\,m^{-1}}$ is the vacuum permittivity. Thus, our strategy is to calculate $\mathcal{P}$, which will then allow us to find χ. Furthermore, since $\mathcal{P} = \sum_{i=0}^{N} \mu_i = N\overline{\langle \mu \rangle}$, where N is the number of atoms and $\mu_i = -er_i$ is the dipole moment of the i-th individual atom, our first goal is to calculate $\overline{\langle \mu \rangle}$.

We use the eigenstates $|1\rangle$ and $|2\rangle$ of the unperturbed Hamiltonian, $\hat{H}_0$, to express operators in a matrix representation. For example, the matrix $\mathbf{H_0}$ is

$$\mathbf{H_0} = \begin{bmatrix} \langle 1|\hat{H}_0|1\rangle & \langle 1|\hat{H}_0|2\rangle \\ \langle 2|\hat{H}_0|1\rangle & \langle 2|\hat{H}_0|2\rangle \end{bmatrix} = \begin{bmatrix} E_1 & 0 \\ 0 & E_2 \end{bmatrix}. \tag{6.61}$$

Assuming that the light field, $\mathcal{E}(t)$, is polarized in the x-direction, the matrix representation of the perturbation Hamiltonian $\hat{W}$ due to electric dipole light–matter coupling is

$$\mathbf{W} = \begin{bmatrix} \langle 1|\hat{W}|1\rangle & \langle 1|\hat{W}|2\rangle \\ \langle 2|\hat{W}|1\rangle & \langle 2|\hat{W}|2\rangle \end{bmatrix} = \begin{bmatrix} 0 & -\mu\mathcal{E}(t) \\ -\mu\mathcal{E}(t) & 0 \end{bmatrix}, \tag{6.62}$$

where $\mu := \mu_{12} = \mu_{21}$, $\mu_{ij} = e\langle i|x|j\rangle$, and we assume that each eigenstate has definite parity such that $\mu_{11} = \mu_{22} = 0$. The matrix representation of the total Hamiltonian, $\hat{H} = \hat{H}_0 + \hat{W}$, is then

$$\mathbf{H} = \mathbf{H_0} + \mathbf{W} = \begin{bmatrix} E_1 & -\mu\mathcal{E}(t) \\ -\mu\mathcal{E}(t) & E_2 \end{bmatrix}. \tag{6.63}$$

Given these matrix elements, the ensemble average of the expectation value of the electric dipole moment can be related to the off-diagonal elements of the density matrix using Equation (6.56):[22]

[22] Here, the hermiticity of the density matrix, $\rho_{12} = \rho_{21}^*$, was used at the last step.

$$\overline{\langle\mu\rangle} = \text{Tr}(\mu\rho) = \sum_{m,n} \mu_{mn}\rho_{nm}$$

$$= \mu_{11}\rho_{11} + \mu_{12}\rho_{21} + \mu_{21}\rho_{12} + \mu_{22}\rho_{22} = \mu(\rho_{21} + \rho_{12}) = \mu(\rho_{21} + \rho_{21}^{*}). \tag{6.64}$$

Let us solve the density matrix equation of motion [Equation (6.58)] for our system of resonantly excited two-level atoms. By putting in the matrix elements of ρ and $\mathbf{H}$, we can write

$$\frac{d}{dt}\begin{bmatrix} \rho_{11} & \rho_{21}^{*} \\ \rho_{21} & \rho_{22} \end{bmatrix}$$

$$= \frac{1}{i\hbar}\left(\begin{bmatrix} E_1 & -\mu\mathcal{E} \\ -\mu\mathcal{E} & E_2 \end{bmatrix}\begin{bmatrix} \rho_{11} & \rho_{21}^{*} \\ \rho_{21} & \rho_{22} \end{bmatrix} - \begin{bmatrix} \rho_{11} & \rho_{21}^{*} \\ \rho_{21} & \rho_{22} \end{bmatrix}\begin{bmatrix} E_1 & -\mu\mathcal{E} \\ -\mu\mathcal{E} & E_2 \end{bmatrix}\right), \tag{6.65}$$

from which we get

$$\frac{d\rho_{11}}{dt} = \frac{i\mu\mathcal{E}(t)}{\hbar}(\rho_{21} - \rho_{21}^{*}), \tag{6.66}$$

$$\frac{d\rho_{22}}{dt} = -\frac{i\mu\mathcal{E}(t)}{\hbar}(\rho_{21} - \rho_{21}^{*}). \tag{6.67}$$

By combining these two equations, we obtain

$$\frac{d}{dt}(\rho_{11} - \rho_{22}) = \frac{2i\mu\mathcal{E}(t)}{\hbar}(\rho_{21} - \rho_{21}^{*}). \tag{6.68}$$

Similarly, for the time derivative of ρ_{21}, we get

$$\frac{d\rho_{21}}{dt} = -i\omega_0\rho_{21} + \frac{i\mu\mathcal{E}(t)}{\hbar}(\rho_{11} - \rho_{22}). \tag{6.69}$$

6.4.1   Relaxation Times: T_1 and T_2

Let N be the total number of atoms in the two-level ensemble, i.e., $N = N_1 + N_2$, where N_1 (N_2) is the number of atoms in the state $|1\rangle$ ($|2\rangle$). On the other hand, $\rho_{11} = \overline{|c_1|}^2 = N_1/N$ and $\rho_{22} = \overline{|c_1|}^2 = N_2/N$, in accordance with $\text{Tr}(\rho) = \rho_{11} + \rho_{22} = 1$. The population difference, $\Delta N = N_1 - N_2$, can then be related to $\rho_{11} - \rho_{22}$ as

$$\frac{\Delta N}{N} = \frac{N_1 - N_2}{N} = \rho_{11} - \rho_{22}. \tag{6.70}$$

When $\mathcal{E}(t) = 0$, ΔN is given by the thermal equilibrium value, ΔN^{eq}, determined by the Boltzmann factor, i.e.,

$$N_2^{\text{eq}} = N_1^{\text{eq}} e^{-\Theta}, \tag{6.71}$$

where

$$\Theta := \frac{E_2 - E_1}{k_B T} = \frac{\hbar \omega_{21}}{k_B T} = \frac{\hbar \omega_0}{k_B T}. \tag{6.72}$$

Then, since $N = N_1^{eq} + N_2^{eq} = N_1^{eq}(1 + e^{-\Theta})$, we deduce that[23]

[23] From these results, one can see that, as $T \to 0$, $N_1^{eq}/N \to 1$ and $N_2^{eq}/N \to 0$. So, all the atoms will fall down to the lower state, $|1\rangle$, as T approaches zero. Also, as $T \to \infty$, $N_1^{eq}/N \to 0.5^+$ and $N_2^{eq}/N \to 0.5^-$. So, no matter how high the temperature gets, $\Delta N^{eq} > 0$, i.e., no population inversion occurs.

$$\frac{N_1^{eq}}{N} = \frac{1}{1 + e^{-\Theta}}, \tag{6.73}$$

$$\frac{N_2^{eq}}{N} = \frac{e^{-\Theta}}{1 + e^{-\Theta}}, \tag{6.74}$$

$$\frac{\Delta N^{eq}}{N} = \frac{1 - e^{-\Theta}}{1 + e^{-\Theta}}. \tag{6.75}$$

When $\mathcal{E}(t)$ is turned off, ΔN relaxes back to ΔN^{eq} with a lifetime of T_1, the population lifetime.[24] With these definitions, we can rewrite Equation (6.68) as

[24] Also known as the longitudinal or spin–lattice relaxation time in the magnetic resonance context.

$$\frac{d}{dt}\left(\frac{\Delta N}{N}\right) = \frac{2i\mu\mathcal{E}(t)}{\hbar}(\rho_{21} - \rho_{21}^*) - \frac{\Delta N - \Delta N^{eq}}{N}\frac{1}{T_1}. \tag{6.76}$$

Similarly, we introduce T_2, the decoherence time,[25] as the characteristic lifetime with which ρ_{21} relaxes back to its equilibrium value when $\mathcal{E}(t)$ is turned off. However, decoherence processes totally destroy superposition states, i.e., the equilibrium value of ρ_{21} is zero. Therefore, we can rewrite Equation (6.69) as

[25] Also known as the transverse or spin–spin relaxation time.

$$\frac{d\rho_{21}}{dt} = -i\omega_0\rho_{21} + \frac{i\mu\mathcal{E}(t)}{\hbar}\frac{\Delta N}{N} - \frac{\rho_{21}}{T_2}. \tag{6.77}$$

6.4.2 Rotating Wave Approximation

When $\mathcal{E}(t) = 0$, the solution for Equation (6.77) is

$$\rho_{21}(t) \propto e^{-i\omega_0 t}e^{-t/T_2}. \tag{6.78}$$

Thus, when $\mathcal{E}(t) \propto \sin\omega t \propto (e^{i\omega t} - e^{-i\omega t})$ [see Equation (6.39)], the second term ($\propto e^{-i\omega t}$) mainly interacts with the system resonantly since $\omega \approx \omega_0$. Hence, we can neglect the first term ($\propto e^{i\omega t}$).[26] We introduce a slowly varying variable $\sigma_{21}(t)$ and its complex conjugate $\sigma_{21}^*(t) = \sigma_{12}(t)$ through

[26] In this approximation, i.e., the rotating wave approximation, we are rotating together with the $e^{-i\omega t}$ field and monitoring the dynamics of the driven system, which should look slow in this new reference frame.

$$\rho_{21}(t) = \sigma_{21}(t)\, e^{-i\omega t}. \tag{6.79}$$

By substituting Equation (6.79) into Equation (6.77) and keeping only those terms that are resonant, we obtain

$$\frac{d\sigma_{21}}{dt} = i\,\delta\omega\,\sigma_{21} + i\frac{\Omega_0}{2}\frac{\Delta N}{N} - \frac{\sigma_{21}}{T_2}, \tag{6.80}$$

where

$$\delta\omega := \omega - \omega_0 \qquad (6.81)$$

is called the detuning, and

$$\Omega_0 := \frac{\mu E_0}{\hbar} \qquad (6.82)$$

is the *on-resonance* Rabi frequency.[27]

Also, Equation (6.76) can now be rewritten as

$$\boxed{\frac{d}{dt}\left(\frac{\Delta N}{N}\right) = -2\Omega_0\,\mathrm{Im}(\sigma_{21}) - \frac{\Delta N - \Delta N^{\mathrm{eq}}}{N}\frac{1}{T_1}.} \qquad (6.83)$$

[27] See Equation (6.13) for the generalized Rabi frequency.

Equations (6.80) and (6.83), together, constitute the *optical* Bloch equations.[28] It should be noted that Equation (6.80) is a complex equation and thus has two independent equations corresponding to the real and imaginary parts; hence, three coupled differential equations comprise the optical Bloch equations, as in the original Bloch equations developed for magnetic resonance.

[28] The original Bloch equations for magnetic resonance were derived by Felix Bloch [*Physical Review* **70**, 4604 (1946)]. Extensions to the optical Bloch equations were produced by Richard Feynman, Frank Vernon, and Robert Hellwarth [*Journal of Applied Physics* **28**, 49 (1957)] as well as by Fortunato Tito Arecchi and Rodolfo Bonifacio, *IEEE Journal of Quantum Electronics* **1**, 169 (1965)].

6.4.3   Steady-State Solutions

Let us seek steady-state solutions for the optical Bloch equations, as is appropriate for continuous-wave excitation. By setting the time derivative to zero in Equations (6.80) and (6.83), we obtain the following three coupled equations to be solved for $\mathrm{Re}(\sigma_{21})$, $\mathrm{Im}(\sigma_{21})$, and $\Delta N/N$:

$$\mathrm{Re}(\sigma_{21}) = -\delta\omega\,T_2\,\mathrm{Im}(\sigma_{21}), \qquad (6.84)$$

$$\mathrm{Im}(\sigma_{21}) = \delta\omega\,T_2\,\mathrm{Re}(\sigma_{21}) + \frac{\Omega_0 T_2}{2}\frac{\Delta N}{N}, \qquad (6.85)$$

$$\frac{\Delta N}{N} = \frac{\Delta N^{\mathrm{eq}}}{N} - 2\Omega_0\,T_1\,\mathrm{Im}(\sigma_{21}). \qquad (6.86)$$

The solutions are

$$\mathrm{Im}(\sigma_{21}) = \frac{(\Omega_0 T_2)/2}{1 + (\delta\omega)^2 T_2^2 + \Omega_0^2 T_1 T_2}\frac{\Delta N^{\mathrm{eq}}}{N}, \qquad (6.87)$$

$$\mathrm{Re}(\sigma_{21}) = \frac{-\delta\omega\,\Omega_0\,T_2^2/2}{1 + (\delta\omega)^2 T_2^2 + \Omega_0^2 T_1 T_2}\frac{\Delta N^{\mathrm{eq}}}{N}, \qquad (6.88)$$

$$\frac{\Delta N}{N} = \frac{1 + (\delta\omega)^2 T_2^2}{1 + (\delta\omega)^2 T_2^2 + \Omega_0^2 T_1 T_2}\frac{\Delta N^{\mathrm{eq}}}{N}. \qquad (6.89)$$

On dividing Equation (6.87) by Equation (6.89), we can write

$$\mathrm{Im}(\sigma_{21}) = \frac{\pi\Omega_0 T_2}{2}\frac{\Delta N}{N}g_0(\omega), \qquad (6.90)$$

where

$$g_0(\omega) = \frac{1}{\pi} \frac{1}{1 + (\delta\omega)^2 T_2^2} = \frac{1}{\pi} \frac{(\Delta\omega_0/2)^2}{(\delta\omega)^2 + (\Delta\omega_0/2)^2} \tag{6.91}$$

is a normalized Lorentzian with a full width at half maximum (FWHM) linewidth given by

$$\Delta\omega_0 := \frac{2}{T_2}, \tag{6.92}$$

and $\int_{-\infty}^{\infty} g_0(\omega)d\omega = 1$. Similarly, the real part of σ_{21} can be written in compact form:

$$\text{Re}(\sigma_{21}) = -\frac{\pi\Omega_0 T_2^2}{2} \delta\omega \frac{\Delta N}{N} g_0(\omega). \tag{6.93}$$

Now, the polarization, $\mathcal{P}$, of the atomic system with optical susceptibility $\chi(\omega)$ in response to $\mathcal{E} = \mathcal{E}_0 e^{-i\omega t}$ is given by

$$\begin{aligned}
\mathcal{P} &= \text{Re}(\varepsilon_0 \chi \mathcal{E}_0 e^{-i\omega t}) \\
&= \varepsilon_0 \mathcal{E}_0 \text{Re}\left\{(\chi' + i\chi'')(\cos\omega t - i\sin\omega t)\right\} \\
&= \varepsilon_0 \mathcal{E}_0 (\chi' \cos\omega t + \chi'' \sin\omega t).
\end{aligned} \tag{6.94}$$

On the other hand, as we derived via the density matrix formalism,

$$\begin{aligned}
\mathcal{P} &= N\overline{\langle\mu\rangle} = N\mu(\rho_{21} + \rho_{21}^*) \\
&= N\mu \, \text{Re}(\rho_{21}) = 2N\mu \, \text{Re}(\sigma_{21}e^{-i\omega t}) \\
&= 2N\mu \left\{\text{Re}(\sigma_{21})\cos\omega t + \text{Im}(\sigma_{21})\sin\omega t\right\}.
\end{aligned} \tag{6.95}$$

By comparing Equations (6.94) and (6.95), we can identify

$$\chi' = \frac{2N\mu}{\varepsilon_0 E_0} \text{Re}(\sigma_{21}), \tag{6.96}$$

$$\chi'' = \frac{2N\mu}{\varepsilon_0 E_0} \text{Im}(\sigma_{21}). \tag{6.97}$$

[29] It should be noted that both $\delta\omega$ [Equation (6.81)] and ΔN [Equation (6.89)] are functions of ω.

Therefore, using Equations (6.90) and (6.93), we finally obtain the real and imaginary parts of the optical susceptibility:[29]

$$\chi'(\omega) = -\frac{\pi\mu^2}{\varepsilon_0 \hbar} T_2^2 \delta\omega \, \Delta N \, g_0(\omega), \tag{6.98}$$

$$\chi''(\omega) = \frac{\pi\mu^2}{\varepsilon_0 \hbar} T_2 \, \Delta N \, g_0(\omega). \tag{6.99}$$

6.4.4   Excitation-Power-Dependent Linewidth

In the weak excitation limit defined by

$$\Omega_0^2 \, T_1 T_2 \ll 1, \tag{6.100}$$

Equation (6.89) gives $\Delta N \approx \Delta N^{\text{eq}}$, which is independent of ω. Thus, in this limit,

$$\chi'_{\text{weak}}(\omega) \propto (\omega - \omega_0)g_0(\omega), \tag{6.101}$$

$$\chi''_{\text{weak}}(\omega) \propto g_0(\omega). \tag{6.102}$$

However, more generally, the lineshape of $\chi'(\omega)$ and $\chi''(\omega)$ changes with the intensity of the excitation light. This can be readily seen as follows:

$$
\begin{aligned}
\chi''(\omega) &\propto \Delta N g_0(\omega) \\
&\propto \frac{1 + (\delta\omega)^2 T_2^2}{1 + (\delta\omega)^2 T_2^2 + \Omega_0^2 T_1 T_2} \frac{1}{1 + (\delta\omega)^2 T_2^2} \\
&= \frac{(1/T_2)^2}{(\delta\omega)^2 + \left[(1/T_2)^2 + \Omega_0^2 T_1/T_2\right]} := \pi g(\omega).
\end{aligned}
\tag{6.103}
$$

We can further show that the spectral function $g(\omega)$ is a Lorentzian with an excitation-power-dependent FWHM, given by

$$\Delta\omega = 2\sqrt{\left(\frac{1}{T_2}\right)^2 + \frac{\Omega_0 T_1}{T_2}} = \Delta\omega_0\sqrt{1 + \Omega_0^2 T_1 T_2}. \tag{6.104}$$

where $\Delta\omega_0$ [Equation (6.92)] is the weak-excitation FWHM.

6.5 Rate Equations, Gain, and Lasing

As the final subject of this chapter, we introduce a powerful general framework for analyzing the interaction of multilevel matter systems with light, including lasers. The framework is based on the optical Bloch equations and introduces rates to take into account different microscopic dynamic processes, including radiative and nonradiative excitation and relaxation pathways.

6.5.1 Einstein's A and B Coefficients

There are three elementary processes that can occur when a two-level atom interacts with radiation: (a) (stimulated) absorption, (b) stimulated emission, and (c) spontaneous emission. See Figure 6.4. In (a), a photon is absorbed, and the atom is excited from the lower state to the upper state. In (b), a photon deexcites the atom from the upper state to the lower state; as a result, an additional photon is generated. In (c), the atom is initially in the upper state, and it spontaneously decays to the lower state while emitting a photon.

Let us consider N two-level atoms in thermal equilibrium with the walls of a cavity at temperature T. The photon density distribution,

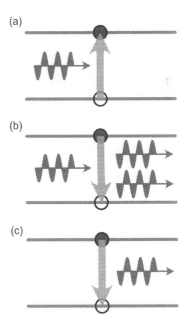

Figure 6.4 The three fundamental processes that occur when a two-level atom resonantly interacts with radiation. (a) (stimulated) absorption, (b) stimulated emission, and (c) spontaneous emission. © Deyin Kong (Rice University).

$\rho_{\mathrm{ph}}(\nu)$, is given by Planck's black-body radiation spectrum:

$$\rho_{\mathrm{ph}}(\nu) = \frac{F(\nu)}{e^{\Theta} - 1}, \tag{6.105}$$

$$F(\nu) := \frac{8\pi h\nu^3}{c^3}, \tag{6.106}$$

where $\Theta := h\nu/k_{\mathrm{B}}T$.[30] The rates of the processes (a), (b), and (c) are respectively given by $B_{12}N_1\rho_{\mathrm{ph}}(\nu)$, $B_{21}N_2\rho_{\mathrm{ph}}(\nu)$, and $A_{21}N_2$, where B_{12}, B_{21}, and A_{21} are coefficients to be determined, N_1 (N_2) is the number of atoms in the lower (upper) state, and $N = N_1 + N_2$. Since, in the steady state, these three rates balance, we obtain

$$B_{12}\rho_{\mathrm{ph}}(\nu) = B_{21}\frac{N_2}{N_1}\rho_{\mathrm{ph}}(\nu) + A_{21}\frac{N_2}{N_1}. \tag{6.107}$$

31 Equation 6.71.

However, since $N_2/N_1 = e^{\Theta}$ in thermal equilibrium,[31] this equation can be rewritten as

$$B_{12}F(\nu) = B_{21}F(\nu)e^{-\Theta} + A_{21}(1 - e^{-\Theta}). \tag{6.108}$$

By equating the temperature-independent terms on both sides of this equation, we get

$$B_{21}F(\nu) = A_{21}. \tag{6.109}$$

By equating the temperature-dependent terms, we get

$$B_{12}F(\nu) = A_{21}. \tag{6.110}$$

Therefore, we can conclude that

$$B_{12} = B_{21} := B, \tag{6.111}$$

$$A_{21} := A = B\frac{8\pi h\nu^3}{c^3}. \tag{6.112}$$

These are Einstein's A and B coefficients.

6.5.2   Deriving the B Coefficient Using Density Matrix Equations

We start with the optical Bloch equations under the rotating wave approximation, i.e., Equations (6.80) and (6.83). We first assume that $T_1 \gg T_2$ and $\dot{\sigma} = 0$.[32] The second condition, together with Equation (6.83) as applied to a steady state, allows us to obtain

$$\mathrm{Im}(\sigma_{21}) = \frac{(\Omega T_2)/2}{1 + (\delta\omega)^2\, T_2^2}\frac{\Delta N}{N}. \tag{6.113}$$

Substitution of this into Equation (6.80) under $T_1 \gg T_2$ leads to

$$\Delta \dot{N} = -\frac{\Omega^2 T_2}{1 + (\omega)^2 T_2^2} \Delta N. \qquad (6.114)$$

However, since

$$\Delta \dot{N} = \dot{N}_1 - \dot{N}_2 = \dot{N}_1 - (\dot{N} - \dot{N}_1) = 2\dot{N}_1, \qquad (6.115)$$

we have

$$\Delta \dot{N}_1 = \frac{\Omega^2}{2} \frac{T_2}{1 + (\delta\omega)^2 T_2^2} (N_2 - N_1). \qquad (6.116)$$

On the other hand, we expect that

$$\Delta \dot{N}_1 = R_{21} N_2 - R_{12} N_1, \qquad (6.117)$$

where R_{21} (R_{12}) is the transition rate (in s^{-1}) for the stimulated emission (absorption) process in this two-level system. Comparison of Equations (6.116) and (6.117) leads us to

$$R_{21} = R_{12} := R = \frac{\Omega^2}{2} \frac{T_2}{1 + (\delta\omega)^2 T_2^2}. \qquad (6.118)$$

It is convenient to write

$$R = \sigma_{\text{op}} \Phi, \qquad (6.119)$$

where σ_{op} (given in m^2) is the optical absorption cross-section and Φ (given in $\text{s}^{-1}\text{m}^{-2}$) is the photon flux. Furthermore, the photon flux can be expressed in terms of the electric field amplitude $\mathcal{E}_0$ as

$$\Phi = \frac{1}{2} \frac{n_{\text{op}} c \varepsilon_0 \mathcal{E}_0^2}{\hbar \omega}. \qquad (6.120)$$

Therefore, by combining these equations, we obtain

$$\sigma_{\text{op}} = \frac{R}{\Phi} = \frac{\mu^2 \omega}{\varepsilon_0 \hbar c n_{\text{op}}} \frac{T_2}{1 + (\delta\omega)^2 T_2^2} = \frac{\pi \mu^2 \omega}{\varepsilon \hbar c n_{\text{op}}} g_0(\omega), \qquad (6.121)$$

where $g_0(\omega)$ is the normalized Lorentzian [Equation (6.91)]. We can further calculate the number of photons in a cavity of volume V as

$$n_{\text{ph}} = \frac{\Phi}{c/n_{\text{op}}} V. \qquad (6.122)$$

Hence,

$$R = \sigma_{\text{op}} \frac{n_{\text{ph}} c}{n_{\text{op}} V} = \frac{\sigma_{\text{op}} c}{n_{\text{op}} V} n_{\text{ph}}. \qquad (6.123)$$

This means that each photon has a probability $\sigma_{op}c/n_{op}V$ of inducing a transition. By introducing the energy spectral distribution, $\rho_{ph}(\omega)$,[33] we can calculate the total absorption/stimulated emission rate to be

$$R = \int_0^\infty \frac{\sigma_{op}c}{n_{op}V} \frac{\rho_{ph}(\omega)V}{\hbar\omega} d\omega. \tag{6.124}$$

Approximating $g_0(\omega)$ by a delta function in Equation (6.121), we finally obtain

$$R = \rho_{ph} \frac{\pi\mu^2}{\hbar^2 \, \varepsilon_0 \, n_{op}^2} = \rho_{ph}B. \tag{6.125}$$

Hence,

$$B = \frac{\pi\mu^2}{\hbar^2 \, \varepsilon_0 \, n_{op}^2}, \tag{6.126}$$

$$A = \frac{1}{t_{spon}} = \frac{\mu^2 \, \omega^3 \, n_{op}}{3\pi \, c^3 \, \hbar \, \varepsilon_0}. \tag{6.127}$$

6.5.3 Absorption versus Gain

In terms of frequency $\nu = \omega/2\pi$ Hz (or s^{-1}), Einstein's A coefficient and the absorption cross-section σ_{op} m^2 can be written as

$$A = \frac{1}{t_{spon}} = \frac{8\pi^2\mu^2\nu^3 \, n_{op}}{3c^3 \, \hbar \, \varepsilon_0}, \tag{6.128}$$

$$\sigma_{op}(\nu) = \frac{\lambda^2}{8\pi t_{spon}} g(\nu), \tag{6.129}$$

where $\lambda = \lambda_0/n_{op}$ and $g(\nu)$ is the lineshape function. For a homogeneously broadened line,

$$g(\nu) = g_0(\nu) = \frac{\Delta\nu/2\pi}{(\nu - \nu_0)^2 + (\Delta\nu/2)^2}. \tag{6.130}$$

The absorption coefficient, $\alpha(\nu)$, and the absorption cross-section, $\sigma_{op}(\nu)$, are related to each other through

$$\alpha(\nu) = \sigma_{op}(\nu)(n_1 - n_2), \tag{6.131}$$

where $n_1 := N_1/V$ and $n_2 := N_2/V$. This equation suggests that $\alpha < 0$, or that there is a finite gain, $\gamma := -\alpha > 0$, when $n_2 > n_1$ (population inversion).[34]

An electromagnetic wave propagating in the medium in the z-direction is written as[35]

[34] Note that in thermal equilibrium $n_2 = n_1 \exp(-E_{21}/k_B T)$, and thus, $n_2 < n_1$ at any temperature.

[35] See Appendix B.

$$\mathcal{E}(z,t) = \mathcal{E}_0 e^{i(\omega t - kz)}, \qquad (6.132)$$

where $k = \tilde{N}k_0 = (n_{op} - i\kappa)k_0 = k' - ik''$ is the (complex) wavenumber or propagation constant; $k' = n_{op}k_0 = \omega/c' = 2\pi\nu/c'$, $c' = c/n_{op}$, and $k'' = \kappa k_0 = \kappa\omega/c = \alpha/2$. Thus the electric field amplitude can be written

$$\mathcal{E}(z,t) = \mathcal{E}_0 e^{i(\omega t - n_{op}k_0 z)} e^{-k''z}, \qquad (6.133)$$

and the intensity is given by

$$I(z) = I(0)e^{-2k''z} = I(0)e^{-\alpha z}. \qquad (6.134)$$

Therefore, depending on the sign of α, and hence the sign of $n_2 - n_1$, the wave will be either attenuated ($\alpha > 0$) or amplified ($\alpha < 0$ or $\gamma > 0$).

6.5.4 Gain Spectrum Calculation

Let us consider a generic three-level system ($|0\rangle$, $|1\rangle$, $|2\rangle$; $E_0 < E_1 < E_2$) where we wish to create population inversion between the states $|1\rangle$ and $|2\rangle$. We assume that $E_1 - E_0 \gg k_B T$ and $E_2 - E_0 \gg k_B T$, so that we can neglect any thermal equilibrium population in $|1\rangle$ and $|2\rangle$ (i.e., $n_1^{eq} = n_2^{eq} = 0$). We introduce population relaxation times τ_{21} and τ_{10} to represent relaxation from $|2\rangle$ to $|1\rangle$ and from $|1\rangle$ to $|0\rangle$, respectively. For mathematical simplicity, we neglect direct relaxation from $|2\rangle$ to $|0\rangle$. We define R_2 to be the rate at which the state $|2\rangle$ is excited (or "pumped") by some means (electrical or optical). Let Φ be the photon flux inside the cavity. Then we can set up the following rate equations:

$$\dot{n}_2 = R_2 + \sigma_{op}\Phi(n_1 - n_2) - \frac{n_2}{\tau_{21}}, \qquad (6.135)$$

$$\dot{n}_1 = \sigma_{op}\Phi(n_2 - n_1) + \frac{n_2}{\tau_{21}} - \frac{n_1}{\tau_{10}}. \qquad (6.136)$$

First, let us consider the "cold cavity" situation, where the photon flux in the cavity Φ is small, so that the stimulated emission and absorption terms in the above equations can be neglected. Then, in the steady state, we can write

$$0 = R_2 - \frac{n_2}{\tau_{21}} \;\rightarrow\; n_2 = R_2\tau_{21}, \qquad (6.137)$$

$$0 = \frac{n_2}{\tau_{21}} - \frac{n_1}{\tau_{10}} \;\rightarrow\; n_1 = \frac{\tau_{10}}{\tau_{21}} n_2 = R_2\tau_{10}. \qquad (6.138)$$

So the cold-cavity population difference is given by

$$n_{d0} = n_2 - n_1 = R_2(\tau_{21} - \tau_{10}). \qquad (6.139)$$

Therefore, we conclude that population inversion ($n_{d0} > 0$) exists when $\tau_{21} > \tau_{10}$. The corresponding gain spectrum is given by

$$\gamma_0(\nu) = \sigma_{op}n_{d0} = \frac{\lambda^2}{8\pi t_{spon}}\, g(\nu)\, n_{d0}. \tag{6.140}$$

In a more general case, the photon flux Φ is finite, and the steady-state rate equations become

$$0 = R_2 + \sigma_{op}\Phi n_1 - \left(\sigma_{op}\Phi + \frac{1}{\tau_{21}}\right) n_2, \tag{6.141}$$

$$0 = \left(\sigma_{op}\Phi + \frac{1}{\tau_{21}}\right) n_2 - \left(\sigma_{op}\Phi + \frac{1}{\tau_{10}}\right) n_1, \tag{6.142}$$

from which we can derive

$$n_d = n_2 - n_1 = \frac{n_{d0}}{1 + \tau_{21}\sigma_{op}\Phi} = \frac{n_{d0}}{1 + \Phi/\Phi_{sat}}, \tag{6.143}$$

where

$$\Phi_{sat} := \frac{1}{\sigma_{op}\tau_{21}} = \frac{8\pi}{\lambda^2} \frac{t_{spon}}{\tau_{21}} \frac{1}{g(\nu)}. \tag{6.144}$$

The gain spectrum in this general case is thus given by

$$\gamma(\nu) = \sigma_{op}(\nu)n_d = \frac{\gamma_0(\nu)}{1 + \Phi/\Phi_{sat}}, \tag{6.145}$$

6.5.5   Two-, Three-, and Four-Level Systems

In order to explore what kind of systems are suited for creating population inversion (and thus gain and lasing), here, in this final section of the chapter, we examine generic two-level, three-level, and four-level systems. The reader should refer to Figure 6.5 in the following discussions.

Figure 6.5(a) shows a two-level system that is resonantly excited through the $|1\rangle$ to $|2\rangle$ transition with photon flux Φ. In terms of the optical cross-section σ_{op} of this transition, we can write the excitation rate as $\sigma_{op}\Phi\, n_1$. At the same time, the state $|2\rangle$ is being depopulated through stimulated emission at rate $\sigma_{op}\Phi\, n_2$. In addition, we assume a spontaneous decay (either radiative or nonradiative) from $|2\rangle$ to $|1\rangle$ with a lifetime of τ_{21}. Thus, the rate equation is given by

$$\dot{n}_2 = \sigma_{op}\Phi\, n_1 - \sigma_{op}\Phi\, n_2 - \frac{n_2}{\tau_{21}}. \tag{6.146}$$

At steady state ($\dot{n}_2 = 0$), this equation leads to

$$n_1 = \left(1 + \frac{1}{\sigma_{\text{op}}\tau_{21}}\right) n_2 = \left(1 + \frac{1}{\eta}\right) n_2, \qquad (6.147)$$

where $\eta := \Phi/\Phi_{\text{sat}}$.[36] Since $n = n_1 + n_2$, we obtain

$$n_1 = \frac{1+\eta}{1+2\eta} n, \qquad (6.148)$$

$$n_2 = \frac{\eta}{1+2\eta} n, \qquad (6.149)$$

$$n_{\text{d}} = n_2 - n_1 = -\frac{1}{1+2\eta} n. \qquad (6.150)$$

Hence, n_{d} is always negative, showing that population inversion is impossible no matter how large Φ is. As $\Phi \to \infty$ (and thus $\eta \to \infty$), n_{d} increases toward zero but never becomes positive.

In essence, the impossibility of achieving population inversion in a two-level system is due to the simultaneous increase in the depopulation rate through stimulated emission as η increases. In other words, the pump photon flux is increasing and removing n_2 at the same time. To circumvent this issue, we place a third level, $|3\rangle$, and assume a fast relaxation process from $|3\rangle$ to $|2\rangle$ with a short lifetime, $\tau_{32} \ll \tau_{21}$, as shown in Figure 6.5(b). Then we can neglect the stimulated emission process $[\sigma_{\text{op}}\Phi (n_1 - n_3) \approx \sigma_{\text{op}}\Phi n_1]$ and write

$$\dot{n}_3 = \sigma_{\text{op}}\Phi n_1 - \frac{n_3}{\tau_{32}}, \qquad (6.151)$$

$$\dot{n}_2 = \frac{n_3}{\tau_{32}} - \frac{n_2}{\tau_{21}}, \qquad (6.152)$$

$$\dot{n}_1 = -\sigma_{\text{op}}\Phi n_1 + \frac{n_2}{\tau_{21}}. \qquad (6.153)$$

By solving these coupled equations in the steady state, we get

$$n_{\text{d}} = n_2 - n_1 = \frac{\eta - 1}{\eta + 1} n. \qquad (6.154)$$

This equation tells us that population inversion ($n_{\text{d}} > 0$) is realized when $\eta > 1$.

Finally, let us consider creating population inversion in the four-level system depicted in Figure 6.5(c). Here, in order to efficiently create population inversion between states $|2\rangle$ and $|1\rangle$, the microscopic processes are designed in such a way that the relaxation from $|2\rangle$ to $|1\rangle$ is the slowest, i.e., $\tau_{21} \gg \tau_{32}, \tau_{10}$. The $|0\rangle$ to $|3\rangle$ transition is constantly pumped at a rate of $\sigma_{\text{op}}\Phi(n_0 - n_3) \approx \sigma_{\text{op}}\Phi n_0$. Under these conditions, we can write $n_1 \approx n_3 \approx 0$, $n = n_0 + n_2$, and

$$n_0 = \frac{1}{\sigma_{\text{op}}\Phi\tau_{21}} n_2 = \frac{1}{\eta} n_2. \qquad (6.155)$$

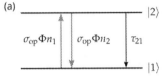

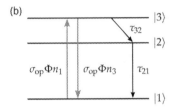

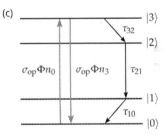

Figure 6.5 Rate equation analysis of pumped (a) two-level, (b) three-level, and (c) four-level systems. © Deyin Kong (Rice University).

Thus, we get

$$n_{\mathrm{d}} = n_2 - n_1 = \frac{\eta}{1+\eta}n. \qquad (6.156)$$

We can see that in this case n_{d} is positive as long as η is finite. Hence, population inversion exists even with infinitesimally weak pumping.

Figure 6.6 Results of rate equation analysis of pumped two-level, three-level, and four-level systems. In a two-level system, population inversion ($n_{\mathrm{d}} > 0$) never occurs. In a three-level system, population inversion can occur above a threshold pump photon flux ($\eta = \Phi/\Phi_{\mathrm{sat}} = 1$), while it always occurs in a four-level system whenever the pump flux is finite. © Deyin Kong (Rice University).

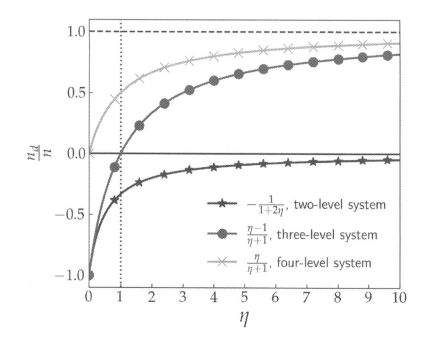

Figure 6.6 plots n_{d}/n as a function of η, based on Equations (6.150), (6.154), and (6.156). It is seen that population inversion is most easily achieved in the four-level system since it occurs whenever η is finite. In the three-level system, it is possible to achieve population inversion if one pumps hard enough. In the two-level system, no population inversion occurs no matter how hard one pumps.

6.6   Chapter Summary

In this chapter, we surveyed the basic quantum mechanical formalism for describing light–matter interaction phenomena. This is an important topic especially as it has applications in semiconductor optoelectronic devices, such as LEDs, lasers, photodetectors, and solar cells. We started from the fundamental Schrödinger equation for a multilevel system resonantly interacting with an electromagnetic field and analyzed the time evolution of the system using different approaches. Using perturbation theory we were able to derive Fermi's golden rule, which allows one to calculate transition probabilities. Using the density matrix formalism, we derived the optical Bloch equations within the rotating wave approximation, with which we deduced an expression for the optical susceptibility for a two-level system. Further, by considering the three fundamental microscopic processes – absorption, stimulated emission, and spontaneous emission – in multilevel systems, we derived the conditions under which a system exhibits finite gain and amplification.

6.7 Exercises

Exercise 6.1 (Classical Lorentz model)

Let us consider the classical Lorentz model for light–atom interaction, where an electron of charge $-e$ and mass m_e is connected to a spring of constant κ_0 and driven by an AC electric field, $\mathcal{E} = \mathcal{E}_0 \cos \omega t$, as shown in Figure 6.7.

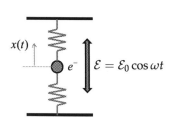

(a) Set up an equation of motion for the displacement $x(t)$.

Figure 6.7 The classical Lorentz model of light–atom interaction.

(b) For a collection of N atoms, the total polarization $\mathcal{P}$ is equal to $N\mu$, where $\mu = -ex(t)$ is the dipole moment of each atom. Derive an equation of motion for $\mathcal{P}$, including the damping term $-2\gamma\, d\mathcal{P}/dt$, where γ is the damping rate.

(c) Assume a linear response, i.e., $\mathcal{E} \sim e^{i\omega t}$ and $\mathcal{P} \sim e^{i\omega t}$, and obtain the frequency dependence of the optical susceptibility $\chi = \mathcal{P}/\varepsilon_0\mathcal{E}$. Use $\omega_0 = \sqrt{\kappa_0/m_e}$.

[37] This corresponds to the rotating-wave approximation in the quantum case. See Section 6.4.2.

(d) Under the resonance approximation $\omega \approx \omega_0$,[37] derive the following expression for the susceptibility:

$$\chi(\omega) \approx \frac{Ne^2}{2\varepsilon_0 m_e \omega_0} \frac{1}{(\omega_0 - \omega) + i\gamma}. \tag{6.157}$$

Exercise 6.2 (Transmission through a slab)

Let us consider a simple situation where a plane wave propagating in the positive z-direction with angular frequency ω, $\mathcal{E}^{(i)} = \mathcal{E}_0 \exp(ik_0 z - i\omega t)$ (where $k_0 = \omega/c$ is the wavenumber in vacuum), is incident normally on a plane-parallel slab bounded by infinite planes $z = 0$ and $z = l$, as shown in Figure 6.8.

Figure 6.8 Transmission and reflection at a slab of thickness l. © Deyin Kong (Rice University).

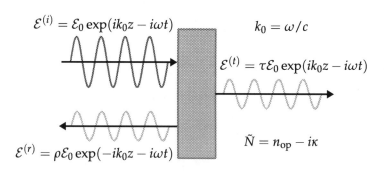

$$\mathcal{E}^{(i)} = \mathcal{E}_0 \exp(ik_0 z - i\omega t) \qquad k_0 = \omega/c$$

$$\mathcal{E}^{(t)} = \tau \mathcal{E}_0 \exp(ik_0 z - i\omega t)$$

$$\mathcal{E}^{(r)} = \rho \mathcal{E}_0 \exp(-ik_0 z - i\omega t)$$

$$\tilde{N} = n_{\mathrm{op}} - i\kappa$$

(a) Including multiple reflections, show that the (complex) reflection and transmission coefficients are expressed, respectively, as

$$\rho = \frac{\mathcal{E}^{(r)}}{\mathcal{E}^{(i)}} = \frac{r\{1 - \exp(2ikl)\}}{1 - r^2 \exp(2ikl)}, \tag{6.158}$$

$$\tau = \frac{\mathcal{E}^{(t)}}{\mathcal{E}^{(i)}} = \frac{(1 - r^2)\exp(ikl)}{1 - r^2 \exp(2ikl)}, \tag{6.159}$$

where

$$r = \frac{1 - \tilde{N}}{1 + \tilde{N}} = \frac{1 - (n_{op} - i\kappa)}{1 + (n_{op} - i\kappa)} \tag{6.160}$$

is the single-surface reflection coefficient, $\tilde{N}$ is the complex refractive index of the slab, and $k = \tilde{N}k_0 = k' - ik''$ ($k_0 = \omega/c$) is the propagation constant inside the slab.

(b) Show that, from Equation (6.158), the fraction of the incident power that is reflected from the slab, called the "slab reflectance" R_s (as opposed to the half-space reflectivity $R = |r|^2$), is given by

$$R_s = \rho\rho^*$$
$$= |r|^2 \left[\frac{(1 - e^{-2k''l})^2 + 4e^{-2k''l}\sin^2(k'l)}{(1 - |r|^2 e^{-2k''l})^2 + 4|r|^2 e^{-2k''l}\sin^2(k'l + \phi_r)} \right], \tag{6.161}$$

and, from Equation (6.159), the transmitted fraction of the power, the "slab transmittance" T_s, is given by

$$T_s = \tau\tau^*$$
$$= e^{-2k''l} \left[\frac{(1 - |r|^2)^2 + 4|r|^2 \sin^2 \phi_r}{(1 - |r|^2 e^{-2k''l})^2 + 4|r|^2 e^{-2k''l}\sin^2(k'l + \phi_r)} \right], \tag{6.162}$$

where $\phi_r = 2\kappa/(1 - n_{op}^2 - \kappa^2)$.

(c) In the case of highly monochromatic waves and low damping, the quantities T_s and R_s display an oscillatory Fabry–Pérot behavior as a function of l, ω, or other parameters affecting k'. This can be most clearly demonstrated by neglecting losses altogether. Derive the expression for T_s in this limit (the lossless limit). What is the period of oscillation?

(d) In the limit where $k'l \gg k''l \gg 1$, show that

$$T_s = (1 - |r|^2)^2 \exp(-2k''l). \tag{6.163}$$

Discuss the physical meaning of the conditions $k'l \gg k''l \gg 1$ as well as each factor appearing in the expression of T_s.

Exercise 6.3 (Transition probability)

A particle is in the ground state in an infinite potential well whose walls are located at $x = 0$ and $x = L$.[38] At $t = 0$, a time-dependent perturbation $\hat{W}(t) = \varepsilon x e^{-t^2}$ is turned on, where ε is a small real number. Calculate the probability that the particle will be found in the first excited state after a sufficiently long time (i.e., $t \to \infty$).

[38] Figure 2.4. See also Equation (2.43).

Exercise 6.4 (Time evolution of a two-level system)

Consider a 2D Hilbert space whose basis states are $\{|0\rangle, |1\rangle\}$. The Hamiltonian operator in this space is given by $\hat{H} = \hbar \omega \sigma_x$ where ω is a real number having the dimension of angular frequency and σ_x is the x-component of the Pauli matrices.

(a) Show that[39]

[39] See Equation (2.81).

$$e^{-i\hat{H}t/\hbar} = \hat{U}(t) = \begin{bmatrix} \cos\omega t & -i\sin\omega t \\ -i\sin\omega t & \cos\omega t \end{bmatrix}. \quad (6.164)$$

(b) Find the solution of the time-dependent Schrödinger equation[40] with the initial condition

[40] Equation (2.79).

$$|\psi(0)\rangle = |0\rangle = \begin{bmatrix} 1 \\ 0 \end{bmatrix}. \quad (6.165)$$

(c) Calculate and discuss the probability $|\langle \psi(0)|\psi(t)\rangle|^2$.

(d) In the Heisenberg picture, operators evolve in time while quantum states are constant. The solution of the Heisenberg equation of motion for the z-component of the Pauli matrices, σ_z,

$$i\hbar \frac{d\sigma_z}{dt} = [\sigma_z, \hat{H}] \quad (6.166)$$

is given by $\sigma_z(t) = e^{i\hat{H}t/\hbar} \sigma_z e^{-i\hat{H}t/\hbar}$. Calculate $\sigma_z(t)$.

(e) Demonstrate the equivalence between the Heisenberg picture and the Schrödingier picture by showing that

$$\langle \psi(0)|\sigma_z(t)|\psi(0)\rangle = \langle \psi(t)|\sigma_z|\psi(t)\rangle. \quad (6.167)$$

Exercise 6.5 (Oscillator strength)

(a) For an infinite-barrier square quantum well,[41] calculate the oscillator strengths F_{00}, F_{01}, and F_{11}.

[41] See Figure 2.4 and Equation (2.43).

(b) Show that

$$\sum_f F_{fi} = 1 \quad (6.168)$$

for any initial state $|i\rangle$ and for any Hamiltonian $\hat{H}_0 = \hat{p}_x^2/2m + V(\hat{x})$. This is called the oscillator strength sum rule. *Hint:* Show that, and then use, $[\hat{x}, \hat{H}_0] = i\hbar \hat{p}_x/m$.

(c) Consider a GaAs quantum well with infinite barrier height. Show that the oscillator strength of the transition between the first two states, $|1\rangle$ and $|2\rangle$, in the quantum well, F_{12}^x, is given by $0.96 m_e / m^*$, where m^* is the effective mass of electrons in the conduction band of GaAs. Use $m^* = 0.07 m_e$ to obtain the oscillator strength for this intersubband transition and compare it with some typical values of oscillator strengths of atomic lines.

Exercise 6.6 (Intersubband transition in a quantum well)

An electron of mass m^* is confined inside a 1D well of length L with infinite barrier height.[42] We are interested in using optical transitions between different states (or "subbands") in this quantum well for developing light detection devices.

[42] See Figure 2.4 and Equation (2.43).

(a) Develop the selection rules for optical transitions between different subbands.

(b) For the lowest-energy allowed transition, calculate the transition energy (in eV) when $L = 10$ nm and $m^* = 0.07 m_e$, where m_e is the free electron mass in vacuum. What is the corresponding wavelength of light absorbed through this transition?

(c) For the lowest-energy allowed transition, calculate the dipole moment (in C m) when $L = 10$ nm and $m^* = 0.07 m_e$. Also express it in units D (debyes), where the conversion is $1 \text{ D} = 3.33 \times 10^{-30}$ C m. What is the oscillator strength of this transition?

(d) For a certain application, we want the wavelength of light detected through the lowest-energy allowed transition to be 10.6 µm. To achieve this goal, we can vary the quantum well thickness L, while we have some flexibility to vary m^* by choosing an appropriate semiconductor. If a larger dipole moment is desired, which of the following two semiconductors is to be preferred as the quantum well material – GaAs (for which $m^* = 0.07 m_e$) or InSb (for which $m^* = 0.015 m_e$)? For the preferred material, calculate the required value of L and the resulting dipole moment (in D) and oscillator strength.

Exercise 6.7 (Absorption saturation)

Show that, when $\delta \omega = 0$, the population difference in Equation (6.89) can be written as

$$\Delta N = \Delta N^{eq} \frac{1}{1 + \Omega_0^2 T_1 T_2} = \Delta N^{eq} \frac{1}{1 + I/I_{sat}}, \qquad (6.169)$$

where $I = n_{op}c\varepsilon_0\mathcal{E}_0^2/2$, n_{op} is the real part of the refractive index, and

$$I_{sat} := \frac{\hbar^2 n_{op}c\varepsilon_0}{2\mu^2 T_1 T_2} \tag{6.170}$$

is called the saturation intensity; $\Delta N = 0.5\Delta N^{eq}$ when $I = I_{sat}$.

Exercise 6.8 (Quantum engineering and rate equations)

Figure 6.9 shows the energy levels involved in one unit cell of a simplified quantum cascade laser structure through which electrons propagate from $z = -\infty$ to $z = \infty$, where z is the growth direction (the direction along which the crystal was grown, layer by layer).

Figure 6.9 A schematic diagram of one unit cell of a quantum cascade laser structure, consisting of a GaAs quantum well with Al$_x$Ga$_{1-x}$As barriers ($0 < x \leq 1$). © Deyin Kong (Rice University).

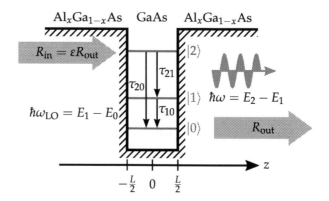

We wish to achieve steady-state population inversion between the states $|2\rangle$ and $|1\rangle$. Here τ_{21}, τ_{20}, and τ_{10} are the relaxation times for electron relaxation from $|2\rangle$ to $|1\rangle$, from $|2\rangle$ to $|0\rangle$, and from $|1\rangle$ to $|0\rangle$, respectively; R_{out} is the rate at which electrons in state $|0\rangle$ tunnel out of the quantum well to populate state $|2\rangle$ of the adjacent well on the right-hand side (downstream); R_{in} ($= \varepsilon R_{out}$) is the rate at which state $|2\rangle$ of the well is populated by electrons coming out of state $|0\rangle$ of the adjacent well on the left-hand side (upstream), with an efficiency ε ($0 \leq \varepsilon \leq 1$). The quantum well material is GaAs, and the barrier material is Al$_x$Ga$_{1-x}$As ($0 < x \leq 1$). Assume that the conduction band offset (i.e., barrier height) $\Delta E_c(x) = 0.3x$ eV and that the electron effective mass $m^*(x) = [0.067(1-x) + 0.15x]m_e$. The well width is L. In order to depopulate state $|1\rangle$ quickly, we utilize resonant longitudinal-optical (LO) phonon scattering, i.e., the energy separation $E_1 - E_0$ needs to be designed to be equal to the LO phonon energy of GaAs, $\hbar\omega_{LO} = 36$ meV.

(a) Set up the Schrödinger equation for the z part of the conduction band wavefunction $\chi(z)$. By imposing appropriate boundary conditions on the wavefunction, derive the following equations for even- and odd-parity bound-state solutions:

$$\nu\,\zeta\tan\zeta = \eta \quad \text{(even)}, \tag{6.171}$$

$$\nu\,\zeta\cot\zeta = -\eta \quad \text{(odd)}, \tag{6.172}$$

where $\nu(x) = m^*(x)/m^*(0)$, $\zeta = \alpha L/2$, $\eta(x) = \beta(x)L/2$, $\alpha = \sqrt{2m^*(0)E}/\hbar$, $\beta(x) = \sqrt{2m^*(x)\left[\Delta E_c(x) - E\right]}/\hbar$, and E is the energy of an electron.

(b) Under the constraint that $E_1 - E_0 = \hbar\omega_{LO} = 36\,\text{meV}$, find values for L and x that make the wavelength of light emitted through the $|2\rangle$ to $|1\rangle$ transition equal to $22.6\,\mu\text{m}$. Restrict the ranges of L and x so that there are exactly three bound states in the quantum well.

(c) Set up rate equations for the three states, including the photon flux Φ in the cavity and defining the absorption cross-section of the $|1\rangle$–$|2\rangle$ transition to be σ_{op}. By setting $d/dt = 0$ and $\Phi = 0$, express the steady-state cold-cavity population inversion $n_{do} = n_2 - n_1$ in terms of τ_{21}, τ_{20}, τ_{10}, R_{out}, and ε. Discuss how to maximize n_{do} when $\varepsilon = 1$ and $\tau_{20} \gg \tau_{21}, \tau_{10}$.

Exercise 6.9 (Temperature dependence)

The band gap of $Al_xGa_{1-x}As$ at $T = 300\,\text{K}$ is given approximately by $E_g(x) = 1.424 + 1.266x + 0.26x^2\,\text{eV}$.

(a) An $Al_xGa_{1-x}As$ LED has an emission peak at $822\,\text{nm}$ at $25\,°\text{C}$. What is the band gap? What is the composition x?

(b) Using the Varshni formula,

$$E_g(T) = E_g(0) - \frac{AT^2}{B + T}, \tag{6.173}$$

find the peak emission wavelength of the emission spectrum from an $In_{0.47}Ga_{0.53}As$ LED when it is cooled from $25\,°\text{C}$ to $-25\,°\text{C}$. The Varshni parameters for this semiconductor are $E_g(0) = 0.850\,\text{eV}$, $A = 4.906 \times 10^{-4}\,\text{eV}\,\text{K}^{-1}$, and $B = 301\,\text{K}$.

(c) Consider a Fabry–Pérot semiconductor laser diode with a cavity length of $250\,\mu\text{m}$ operating at $1550\,\text{nm}$ and $T = 300\,\text{K}$. The refractive index (n_{op}) of the active layer at $300\,\text{K}$ is 3.60, $dn_{op}/dT \approx 2 \times 10^{-4}\,\text{K}^{-1}$, and the linear thermal expansion coefficient is $5.6 \times 10^{-6}\,\text{K}^{-1}$. Find the shift in the emission wavelength for a given mode per unit temperature change.

7

Quantum States in Technological Materials

Learning objectives:

- Developing an intuitive picture of how discrete energy levels become bands when atoms or molecules are brought together to form a crystal.

- Becoming familiar with the concepts of Bragg diffraction and reciprocal lattice.

- Understanding the basic band structures of various semiconductors and quantum semiconductor structures.

- Gaining basic knowledge of the band structures of cutting-edge nanomaterials.

THE BAND THEORY of solids provides a general framework with which to understand properties of materials. It not only explains the fundamental differences in electronic structure between insulators, semiconductors, and metals but also provides guidelines for finding optimum materials for specific device applications. For example, a semiconductor with a light effective mass is suited for high-electron-mobility transistors (HEMTs) because the mobility μ_e is inversely proportional to the effective mass, $\mu_e = e\tau/m^*$, where τ is the scattering time. For developing LEDs and laser diodes, a direct band gap material – i.e., a material in which the conduction-band bottom and the valence-band top occur at the same k – is necessary for momentum conservation since the momentum of photons is negligibly small compared with crystal momenta. In this chapter, after reviewing the basic concepts of atomic and molecular orbitals, bonds and bands, crystal lattices and reciprocal lattices, we provide an overview of the band structure of technologically important materials, including both traditional and emerging materials.

7.1 From Bonds to Bands

As noted in Section 5.4, a crystalline material consists of an array of atoms, ions, or molecules, which creates a spatially periodic potential for electrons.[1] Let us simulate a crystal by placing atoms in proximity with each other so that they are weakly coupled or *bonded* with their neighboring atoms. We assume that the interatomic coupling is weak in the sense that each atom still retains its identity; this is the *tight-binding* picture or model. The periodic potential is then a

[1] Therefore, the electrons obey the Bloch theorem. See Section 5.4.1.

superposition of identical atomic potentials centered at periodically located ion cores with period a, as shown in Figure 7.1.

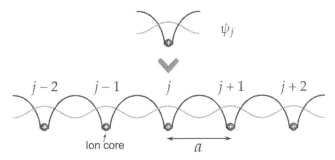

Figure 7.1 The periodic potential (black curves) of a crystalline material can be thought of as a superposition of atomic potentials centered at periodically located ion cores, with period a. The total electron wavefunction $|\Psi\rangle$ is shown in gray. © Deyin Kong (Rice University).

Within an isolated atom, there are multiple bound states, or *orbitals*, tightly bound to this particular atom, but let us focus on one particular orbital, $|\psi\rangle$, for simplicity. The total wavefunction can then be represented by a linear combination of atomic orbitals (LCAO) localized at the atomic sites:

$$|\Psi\rangle = \sum_j c_j(t)|\psi_j(x)\rangle, \tag{7.1}$$

where c_j are time-dependent complex coefficients and $|\psi_j\rangle$ is the atomic orbital localized at the jth atom. Note that $|\Psi(x)\rangle$ extends throughout the crystal.

The time-dependent Schrödinger equation we need to solve for $|\Psi\rangle$ is $\hat{H}|\Psi\rangle = i\hbar\partial|\Psi\rangle/\partial t$, where $\hat{H} = -(\hbar^2/2m)d^2/dx^2 + V(x)$.[2] We substitute the LCAO [Equation (7.1)] for $|\Psi\rangle$ to obtain

$$\sum_j c_j\hat{H}|\psi_j\rangle = i\hbar\sum_j \frac{dc_j}{dt}|\psi_j\rangle. \tag{7.2}$$

[2] $V(x)$ is the superposition of the atomic potentials and is a periodic function, $V(x+a) = V(x)$.

Furthermore, we multiply this equation from the left by the orbital $\langle\psi_k|$, which is localized at the kth atomic site, and integrate both sides. By assuming that the "overlap integrals" between orbitals localized at different atomic sites are negligibly small ($\langle\psi_k|\psi_j\rangle \approx 0$), we derive

$$\sum_j c_j\langle\psi_k|\hat{H}|\psi_j\rangle = i\hbar\sum_j \frac{dc_j}{dt}\langle\psi_k|\psi_j\rangle$$

$$\Rightarrow \sum_j H_{kj}c_j = i\hbar\sum_j \frac{dc_j}{dt}\delta_{jk} = i\hbar\frac{dc_k}{dt}. \tag{7.3}$$

Here, $H_{kj} := \langle\psi_k|\hat{H}|\psi_j\rangle$ is the matrix element of the Hamiltonian between the kth and jth atomic orbitals.

Figure 7.2 When two atoms approach each other closely, their outer orbitals become modified; for each original atomic orbital, two orbitals emerge – the bonding (Ψ_+) and antibonding (Ψ_-) states. They have corresponding eigenenergies $E_\pm = E_0 \mp \Lambda$, where E_0 is the original atomic orbital energy and Λ is the transfer integral. © Deyin Kong (Rice University).

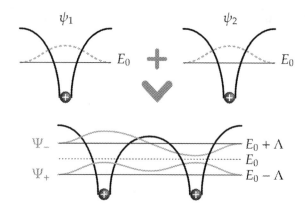

To get a basic understanding of interatomic coupling, let us start with the simplest case, that of two atoms; see Figure 7.2. The LCAO in this case is simply given by $|\Psi\rangle = c_1|\psi_1\rangle + c_2|\psi_2\rangle$, and the coupled differential equations are

$$i\hbar\frac{dc_1}{dt} = H_{11}c_1 + H_{12}c_2, \tag{7.4}$$

$$i\hbar\frac{dc_2}{dt} = H_{21}c_1 + H_{22}c_2. \tag{7.5}$$

Here, because the two atoms are identical, H_{11} must be equal to H_{22}, which we define to be E_0.[3] Similarly, by symmetry, H_{12} must be equal to H_{21}, which we denote by Λ.[4] To obtain the energies of the coupled system as the eigenvalues of the time-independent Schrödinger equation, we assume a sinusoidal time dependence for $c_1 = b_1 e^{-iEt/\hbar}$ and $c_2 = b_2 e^{-iEt/\hbar}$. Here, b_1 and b_2 are numerical coefficients to be determined. By substituting these into the coupled differential equations [Equations (7.4) and (7.5)], we get

$$(E - E_0)b_1 + \Lambda b_2 = 0, \tag{7.6}$$

$$\Lambda b_1 + (E - E_0)b_2 = 0. \tag{7.7}$$

For nontrivial solutions to exist for b_1 and b_2, the determinant of the coefficient matrix has to vanish. We thus obtain the eigenenergies

$$E = E_0 \pm \Lambda. \tag{7.8}$$

Hence, the original atomic energy level splits into two. One is above and the other is below the original energy, E_0. Correspondingly, the coefficients $b_2^{(\pm)} = \pm b_1^{(\pm)}$, and the normalized wavefunctions are

$$|\Psi_\pm\rangle = b_1^{(\pm)}e^{-iE_\pm t/\hbar}|\psi_1\rangle + b_2^{(\pm)}e^{-iE_\pm t/\hbar}|\psi_2\rangle$$
$$= \frac{1}{\sqrt{2}}(|\psi_1\rangle \pm |\psi_2\rangle)e^{-iE_\pm t}. \tag{7.9}$$

The "+" and "−" solutions, $|\Psi_\pm\rangle$, are referred to as the bonding and antibonding states, respectively.

[3] This is essentially the energy of the original uncoupled orbital.

[4] This parameter Λ (> 0) is called the transfer integral and represents the coupling strength between the two atoms. Its magnitude increases with decreasing atomic distance.

We now extend our two-atom analysis to the case of a 1D array of atoms with spacing a. We assume that coupling exists only between neighboring atoms.[5] Under these conditions, the time rate of change of c_j is affected only by $c_j(t)$, $c_{j-1}(t)$, and $c_{j+1}(t)$, i.e.,

$$i\hbar\frac{dc_j}{dt} = E_0 c_j - \Lambda c_{j-1} - \Lambda c_{j+1}. \tag{7.10}$$

Again, assuming the sinusoidal time dependence appropriate for a steady state with energy E, $c_j = b_j e^{-iEt/\hbar}$, we can eliminate the time t. By substitution, we obtain

$$E b_j = E_0 b_j - \Lambda(b_{j-1} + b_{j+1}). \tag{7.11}$$

Furthermore, on the basis of the Bloch theorem,[6] we assume a sinusoidal spatial dependence for b_j and introduce the Bloch index (or Bloch wavevector) k:

$$b_j = b(x_j) = e^{ikx_j} = e^{ikja}. \tag{7.12}$$

By using Equation (7.12) for b_j in Equation (7.11), we obtain the electron energy versus k, or band dispersion, for this system:

$$E(k) = E_0 - 2\Lambda \cos(ka). \tag{7.13}$$

We can see that the width of the band increases with Λ. In the case of a 3D anisotropic crystal, we can extend our analysis to derive three cosine terms with different Λ's for the three orientations:

$$E(\boldsymbol{k}) = E_0 - 2\Lambda_x \cos(k_x a) - 2\Lambda_y \cos(k_y a) - 2\Lambda_z \cos(k_z a). \tag{7.14}$$

Cosine bands are sometimes called tight-binding bands.

Once the band structure of a material has been calculated and all Bloch states have been filled according to Pauli's exclusion principle,[7] one can tell whether the material is a metal or an insulator, using the following criteria. In a metal, the energy of the highest occupied state[8] is inside a band, that is, the highest occupied band is partially filled. In this case, the electrons at the highest occupied states can be excited into a higher state with an infinitesimally small excitation energy. On the other hand, in an insulator, the highest occupied energy level is separated by a finite energy, the band gap energy, from the lowest unoccupied state, so one needs at least the band gap energy to excite an electron. See Figure 7.3.

To summarize the essential points of the tight-binding approach that we discussed in this section, let us look at Figure 7.4. We start with two completely isolated identical atoms and then gradually bring them together. When their atomic orbitals start overlapping

[5] The so-called nearest-neighbor approximation

[6] Section 5.4.1

[7] Which states that no two electrons can occupy exactly the same quantum state.

[8] Known as the Fermi energy or Fermi level.

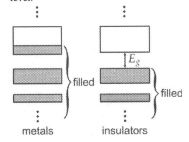

Figure 7.3 Distinction between metals and insulators. © Deyin Kong (Rice University).

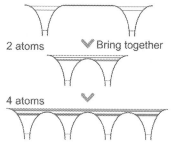

Figure 7.4 From bonds to bands. The more atoms we bring together, the more states that comprise each band. © Deyin Kong (Rice University).

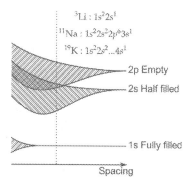

Figure 7.5 A schematic diagram showing how atomic states in an alkaline metal evolve into bands as the interatomic spacing decreases. Since the 1s, 2s, and 2p bands are full, half-filled, and empty, respectively, the Fermi energy resides inside the 2s band, making this system a metal. See Figure 7.3 for the distinction between metals and insulators. The vertical dotted line indicates the equilibrium atomic spacing, which determines the widths of the bands.

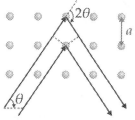

Figure 7.6 When a plane wave is reflected by two adjacent atomic layers, constructive interference occurs when the optical path length $2a\sin\theta$ becomes equal to an integer multiple of the wavelength λ. This is the Bragg condition. © Deyin Kong (Rice University).

each other, we start seeing the splitting of atomic orbitals. Each state splits into two when there are two atoms. When there are four atoms, each atomic energy level splits into four levels that are centered around the original energy. When N atoms are brought together to form a crystal, each atomic energy level splits into N levels, but since N is a large number, these split energy levels form a band. Each atomic level thus forms a band, and the separation between two adjacent bands is the band gap, in which there are no allowed states.

The tight-binding picture is thus powerful and useful for visualizing how atomic states evolve into bands. As an example, let us consider alkali metals, for example lithium, sodium, and potassium. These elements in their atomic form contain an s-orbital as the outermost orbital, which is occupied by one electron. For example, in the case of a lithium atom, the 1s state is fully occupied by two electrons, while the 2s state is occupied only by one electron. Therefore, when N atoms are brought together, the resulting 2s band is exactly half filled, while the 1s band is fully occupied, making this material a metal. See Figure 7.5.

7.2 Bragg Diffraction, Lattice, and Reciprocal Lattice

One very important consequence of the periodic lattice for a propagating wave is Bragg diffraction. This is a general wave phenomenon that occurs for any type of waves, including X-rays, neutron waves, elastic waves, and electron waves. Constructive interference occurs between waves reflected by different atomic layers, whenever the Bragg condition

$$n\lambda = 2a\sin\theta, \quad n = 1, 2, 3, \ldots \quad (7.15)$$

is satisfied. Here, a is the spacing between atomic layers, 2θ is the scattering angle, and λ is the wavelength; see Figure 7.6. The propagating wave will be resonantly scattered into these specific angles for a given a and λ.

Let us now consider a freely propagating wave, $|\psi\rangle \propto e^{ik\cdot r}$, representing a free electron with a parabolic energy–momentum relationship, $E = p^2/2m = \hbar^2 k^2/2m$. For mathematical simplicity, let us focus on a 1D system, for which the Bragg condition becomes particularly simple. The only possible scattering angle in a 1D system is 180°, or $\theta = 90°$, so $\sin\theta = 1$. Namely, in 1D, Bragg diffraction is backscattering. This happens when $n\lambda = 2a$ or $k = 2\pi/\lambda = \pi n/a$, i.e., whenever k is an integer multiple of π/a, where a is the lattice periodicity, or the lattice constant. For example, when $k = +\pi/a$, the electron is resonantly backscattered to the state with $k = -\pi/a$. But the opposite process occurs at the same time, where the electron is scattered from the $k = -\pi/a$ state back to the $k = +\pi/a$ state; see Figure 7.7(a). Therefore, at these special values of k, the wavefunction is a 50%–50% mixture of a wave traveling to the right and a wave traveling to the left, forming a standing wave. There are two ways to form such a standing wave with opposite phases:

$$|\psi_\pm\rangle = \frac{1}{\sqrt{2}}\left[\exp\left(i\frac{\pi x}{a}\right) \pm \exp\left(-i\frac{\pi x}{a}\right)\right]$$

$$= \sqrt{2}\begin{cases}\cos\left(\pi x/a\right), \\ i\sin\left(\pi x/a\right).\end{cases} \tag{7.16}$$

Because these two standing waves, proportional to $\cos\left(\pi x/a\right)$ and to $\sin\left(\pi x/a\right)$ respectively, have large amplitudes at different locations inside the crystal, they have different energies. Therefore, the existence of these two states $|\psi_\pm(x)\rangle$ at these special points in k-space, $\pm\pi/a$, leads to an energy splitting in the originally continuous energy–momentum relationship. See Figure 7.7(b). The first continuum of states is the first band, and is energetically separated from the second band by the energy gap between them. The $k = \pm\pi/a$ states define the first Brillouin zone, as described in Section 5.4.1.

As noted above, a crystalline solid consists of a periodic array of atoms, called a lattice. In a lattice, there is translational symmetry, which means that everything looks exactly the same when you move from one lattice point to another. See Figure 7.8. This is an aspect that we will now explore more rigorously.

Mathematically, a lattice is defined as an infinite set of points defined by integer sums of a set of linearly independent primitive lattice vectors. In 1D, the discrete lattice points are written as $R = na$, where a is the lattice constant and n is an integer. In 2D, lattice points are expressed as lattice vectors given by a linear combination of a_1 and a_2, $R = n_1 a_1 + n_2 a_2$, where n_1 and n_2 are integers and a_1 and a_2 are the primitive lattice vectors. Similarly, in 3D, lattice vectors are given by $R = n_1 a_1 + n_2 a_2 + n_3 a_3$. Thus, when a set of linearly independent vectors a_1, a_2, and a_3 is chosen, the corresponding lattice is defined.[9]

For each valid set of primitive vectors $\{a_i\}$, one obtains a unit cell, which is the elementary building block of the lattice. In other words, one can completely fill the entire space when many such units are stacked together. Another important point to note is that an actual crystal consists of an array of objects, such as atoms, ions, or molecules. Therefore, a crystal is determined not only by the lattice but also by the repeating objects. Each of these repeating objects is called a basis. See Figure 7.9. In a 2D lattice, the controlling parame-

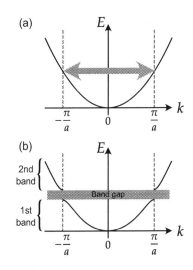

Figure 7.7 (a) In a 1D system, Bragg diffraction is Bragg reflection, which occurs when $k = n\pi/a$. An electron in the $\pm\pi/a$ state is resonantly backscattered into the $\mp\pi/a$ state, resulting in two standing waves. (b) Since the two standing waves have amplitude maxima at different locations in the lattice, they have different energies, leading to an energy gap. © Deyin Kong (Rice University).

[9] However, the converse is not true. For a given lattice, the choice of primitive vectors is not unique.

$$r' = r + R$$
$$R = n_1 a_1 + n_2 a_2 + n_3 a_3$$
$$n_1, n_2, n_3 : \text{integers}$$

Figure 7.8 In a lattice, there is translational symmetry, which means that the atomic arrangement looks exactly the same in every respect when viewed from the point r as when viewed from the point r'. © Deyin Kong (Rice University).

· Lattice point

Repeating object (basis)

Figure 7.9 A crystal is defined by specifying the underlying lattice and the repeating object, called the basis. © Deyin Kong (Rice University).

[10] This kind of data is now called Laue patterns. Different crystals have different symmetries and different periods, so they produce uniquely different Laue patterns.

[11] This form is called the Laue condition, which is fully equivalent to the Bragg condition, Equation (7.15).

ters are the lengths of the primitive lattice vectors $|a_1|$ and $|a_2|$, and the angle between them, φ. Depending on these parameters, there are only five types of lattices in 2D – oblique, rectangular, centered rectangular, hexagonal, and square.

In 3D, there are a total of 14 lattice types – one triclinic lattice, two monoclinic lattices, and four orthorhombic lattices, where $|a_1|$, $|a_2|$, and $|a_3|$ are not equal to each other, but the angles between them are all 90°. There are also two tetragonal lattices, one trigonal lattice, and three cubic lattices, comprising simple cubic, body centered cubic, and face centered cubic. Finally, there is one hexagonal lattice. All these different types of crystal structures have different classes of symmetry, which strongly affect the electronic states and as a result, the electronic, optical, and magnetic properties. Therefore, it is important to characterize the crystal structure of a given material.

In 1912, Max von Laue correctly realized that such a periodic structure is similar to a diffraction grating, so it should diffract light into different directions depending on the wavelength. The only problem is that visible light's wavelengths are very long compared with the spatial period, which corresponds to the wavelengths of X-rays. Therefore, Laue believed that X-ray beams should be diffracted into different directions, and this should tell us about the crystal structure. This was indeed observed: the scattered intensity versus angle of incidence produced a series of peaks.[10] The essence of this phenomenon is Bragg diffraction, shown in Figure 7.6. That is, when the k vector of the incident X-ray is chosen correctly, the scattered radiation from the atoms adds constructively in certain directions. The Bragg condition, $n\lambda = 2a \sin\theta$, can be written in terms of the wavevector k rather than the wavelength λ. Here, we use k' as the wavevector of the scattered wave. We also introduce a vector G as the difference between k' and k, and assume that the scattering is elastic so that the magnitude of k is unchanged. Under these conditions, the Bragg condition can be written as[11]

$$|G| = |k - k'| = \frac{2\pi}{a}n. \tag{7.17}$$

Now we introduce the concept of *reciprocal* space (or k-space), conjugate to the *direct* lattice of a given crystal structure in real space. In 1D we know that each point in the direct lattice is written as na, where a is the period and n is an integer. Namely, $x_n = \ldots, -2a, -a, 0, a, 2a, \ldots$ The corresponding reciprocal lattice has points given by $G_m = 2m\pi/a = \ldots -4\pi/a, -2\pi/a, 0, 2\pi/a, 4\pi/a, \ldots$ Therefore, the period in the reciprocal lattice is $2\pi/a$. With these definitions, one can see that $e^{iG_m x_n}$ is always equal to 1 for any integers m and n. As a result, $e^{i(k+G_m)x_n} = e^{ikx_n}$. So, adding a reciprocal vector to k will not change this quantity, e^{ikx_n}.

In a 3D system, the above condition is written as $e^{iG \cdot R} = 1$. Here, $R = n_1 a_1 + n_2 a_2 + n_3 a_3$ is a direct lattice vector, while the reciprocal lattice vector G is expressed as a linear combination of b_1, b_2, and b_3, which are the primitive lattice vectors of the reciprocal lattice:

$$G = m_1 b_1 + m_2 b_2 + m_3 b_3. \tag{7.18}$$

The b_i's are given by the following expression:

$$b_i = \frac{2\pi a_j \times a_k}{a_i \cdot (a_j \times a_k)}. \tag{7.19}$$

Thus, once we know a_1, a_2, and a_3 for the direct lattice, we can construct the corresponding reciprocal lattice.[12] Further, the primitive lattice vectors of these lattices are related to each other through

[12] One can also prove that the reciprocal of the reciprocal lattice is the original direct lattice itself.

$$a_i \cdot b_j = 2\pi \delta_{ij}. \tag{7.20}$$

One way to look at the direct and reciprocal lattices is that they are related to each other through a Fourier transform. The easiest way to see this is to start with the density of lattice points in 1D, which is written as a series of delta functions,

$$\sum_n \delta(x - na). \tag{7.21}$$

Each delta function is located at each lattice point. The Fourier transform of this series of delta functions is again a series of delta functions, but in k-space (or reciprocal space) instead of real space:

$$\frac{2\pi}{a} \sum_m \delta\left(k - \frac{2\pi m}{a}\right). \tag{7.22}$$

You can see that the period is indeed $2\pi/a$ in the reciprocal lattice.

Let us now go back to the X-ray diffraction problem. As we saw, the Laue condition in 1D is given by $G_m = 2\pi m/a$. In 3D, we can write this condition using vectors, i.e., $G \cdot R = 2\pi m$. However, this is exactly equal to the definition of reciprocal lattice points! Therefore, we can conclude that *every reciprocal lattice point has a corresponding Bragg peak in the Laue pattern*. The reason why a Laue pattern consists of many peaks is that an X-ray can be reflected from different sets of lattice planes, for a given crystal. Because a crystal is a periodic structure, each lattice plane comes with a whole family of lattice planes which are equally spaced in a certain direction. Each family has three integers known as Miller indices, which specify the direction perpendicular to the planes of a given family. See Figure 7.10.

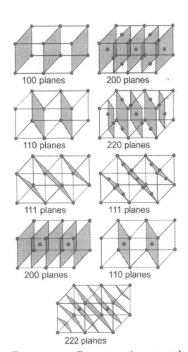

Figure 7.10 Representative crystal planes with their Miller indices. © Deyin Kong (Rice University).

For each family of lattice planes, there is a family of reciprocal lattice vectors that are all perpendicular to the lattice planes. And the shortest vector among them has a length of $2\pi/a$, where a is the lattice spacing. In summary, for a given crystal, there are families of lattice planes having specific Miller indices. Each family shows up as a distinct peak in an X-ray diffraction pattern. Therefore, by analyzing these patterns, one can determine the lattice spacings in different crystal orientations and thus the whole crystal structure.

7.3 Semiconductor Band Structure

The invention of the transistor and the subsequent development of solid-state technology, including computers, proved that semiconductors are useful materials despite earlier negative views.[13] There are many reasons why semiconductors are useful: they possess properties of both metals and insulators, exhibit tunable properties (with temperature, doping, electric field, illumination, etc.), can be n-type or p-type (to make p-n junction diodes and transistors), can produce light with desired wavelengths, and can be processed to fabricate nanostructures with quantum engineered properties.

[13] Wolfgang Pauli is famously quoted to have said, "One shouldn't work on semiconductors, that is a filthy mess; who knows if they really exist!" (Letter to Peierls, September 29, 1931).

(a) (b)

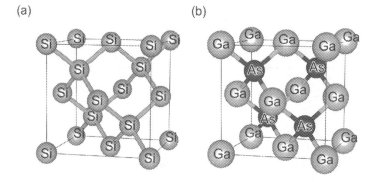

Figure 7.11 Crystal structure of (a) Si and (b) GaAs. These semiconductors are used in traditional electronic and photonic devices such as transistors and laser diodes, respectively. © Deyin Kong (Rice University).

The two most representative semiconductors are Si and GaAs. The former has been the principal material for the development of solid-state information technology, while the latter has revolutionized photonics and optoelectronics. Their crystal structure is shown in Figure 7.11. Silicon crystalizes in the diamond structure, which consists of a pair of intersecting face-centered cubic lattices, separated by 1/4 of the width of the unit cell in each dimension. Gallium arsenide and many other III-V and II-VI compound semiconductors adopt the analogous zinc-blende structure, where the nearest neighbours of each Ga atom are As atoms.

As shown in Figure 7.12(a), the bottom of the conduction band in Si is near the X point, where $(k_x, k_y, k_z) \equiv (\pi/a)(100)$ and a is the lattice constant. The second- and third-lowest-energy points in the conduction band are at the L point $[(k_x, k_y, k_z) \equiv (\pi/a)(111)]$ and the Γ point $[(k_x, k_y, k_z) \equiv (000)]$. On the other hand, the top of the valence band is at the Γ point, so Si is an indirect-gap semiconductor (and so unsuitable for developing light-generation devices, as mentioned above).[14] Note that the valence band arises from p orbitals of Si atoms, which are originally six-fold degenerate. This degeneracy is partially lifted by spin–orbit coupling, leading to the heavy-hole and light-hole bands (which are degenerate at the top of the valence band, i.e., the Γ point) and the spin–orbit split-off band.

As shown in Figures 7.12(b) and 7.12(c), the valence band in Ge and GaAs is essentially the same as that in Si, with the top located at the Γ point. However, the location of the conduction-band bottom differs. In Ge, the conduction-band bottom is near the L point, whereas in GaAs it is at the Γ point. Therefore, Ge is an indirect-gap semiconductor whereas GaAs is a direct-gap semiconductor. The band gap values of GaAs and those of other direct-gap III-V and II-IV semiconductors are summarized in Table 7.1, together with the values of their band-edge electron effective mass[15] and and Kane's interband momentum matrix element.

One of the most important direct-gap semiconductors is gallium nitride (GaN). It has been successfully used in realizing blue LEDs since the 1990s.[16] The compound has a crystal structure known as wurtzite (see Figure 7.13), and its band gap value is 3.4 eV. The development of high-brightness GaN LEDs completed the range of primary colors, enabling novel applications such as daylight-visible full-color LED displays, white LEDs, and blue laser diodes. Combining GaN with other III-nitride compounds – InN and AlN – through the creation of alloys and heterostructures[17] allows the fabrication of LEDs with colors that can go from red to ultraviolet. In addition, GaN is resilient to ionizing radiation, and thus, it is attracting interest as a suitable material for military and space applications. Furthermore, since GaN transistors can operate at high temperatures and high voltages, they are promising for high-power amplifiers at high frequencies.[18]

As an important mechanism in band gap engineering, quantum confinement can change the band gaps of semiconductors, as shown schematically in Figure 2.8. The bottom of the conduction band is pushed up while the top of the valence band is pushed down, and thus, the effective band gap, E_g^*, is larger than the original 3D (or

[14] The wavenumber of a photon with an infrared wavelength, $\lambda = 1\,\mu m$, is $k_{ph} = 2\pi/\lambda \sim 10^5\,cm^{-1}$, whereas the wavenumber of Bloch electrons at the first Brillouin zone edge is $k_{el} = \pi/a \sim 10^9\,cm^{-1}$ with lattice constant $a \sim 0.1\,nm$. Thus, $k_{ph} \ll k_{el}$.

[15] See Equation (5.95).

[16] Isamu Akasaki, Hiroshi Amano, and Shuji Nakamura shared the 2014 Nobel Prize in Physics "For the invention of efficient blue light-emitting diodes, which has enabled bright and energy-saving white light sources."

[17] See Section 7.4.

[18] For a review, see, e.g., U. K. Mishra, L. Shen, T. E. Kazior, and Y. Wu, "GaN-based RF power devices and amplifiers," *Proceedings of the IEEE* **96**, 287 (2008).

(a) Si

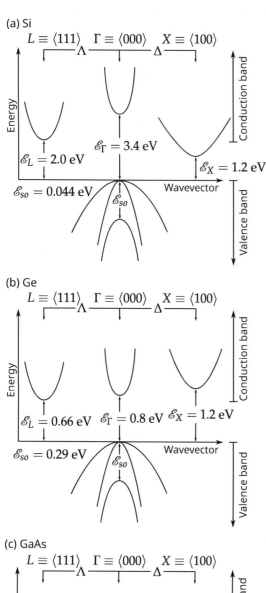

(b) Ge

(c) GaAs

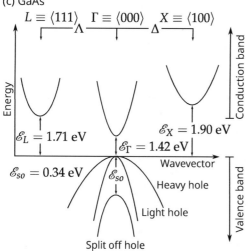

With data from http://www.matprop.ru/

Figure 7.12 Schematic band structure of three typical semiconductors: (a) Si, (b) Ge, and (c) GaAs. The top of the valence band is located at the zone center, the Γ point, (000) in all these semiconductors. The bottom of the conduction band is at the X point, L point, and Γ point, respectively, in Si, Ge, and GaAs. Therefore, GaAs is a direct-gap semiconductor whereas Si and Ge are indirect-gap semiconductors. © Deyin Kong (Rice University).

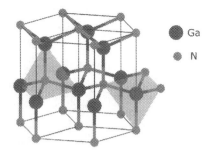

Figure 7.13 The crystal structure of GaN. It has the so-called wurtzite structure. © Deyin Kong (Rice University).

bulk) value, E_g. The smaller the system size becomes, the larger the band gap gets. Figure 7.14 shows schematically a series of aqueous solutions containing semiconductor nanoparticles with progressively different sizes. Because of the size-dependent band gap, different particle sizes lead to different colors.

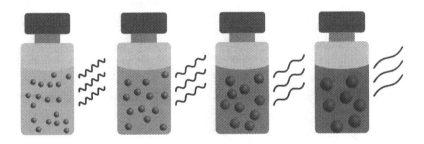

Figure 7.14 Solutions containing nanoparticles made of the same semiconductor material but with different average sizes exhibit different colors due to the size-dependent band gap. © Deyin Kong (Rice University).

There are many different techniques for calculating band structure with varying degrees of approximation. Among them, one particularly useful method for calculating semiconductor band dispersions $E = E(\mathbf{k})$ is called the $\mathbf{k} \cdot \mathbf{p}$ method. This method, originally developed in the 1930s, allows one to obtain $E(\mathbf{k})$ very accurately for near-band-edge states with a set of materials parameters called band parameters.[19] For example, the conduction band dispersion can be written as

$$E_c(\mathbf{k}) = E_c(0) + \frac{\hbar|\mathbf{k}|^2}{2m_e} + \frac{\hbar|\mathbf{k}|^2}{m_e} \sum_{m \neq c} \frac{|\mathbf{k} \cdot \mathbf{p}_{cm}|^2}{E_c(0) - E_m(0)}, \qquad (7.23)$$

where $E_c(0)$ and $E_m(0)$ are band-edge energies of the conduction band and the m-th band, respectively, $\mathbf{p}_{nm} := -i\hbar \langle u_{n0} | \nabla | u_{m0} \rangle$ is the interband momentum matrix element, and m runs through all bands. In the simplest two-band model, where we keep only the valence band ($m = v$) in the summation and $E_g := E_c(0) - E_v(0)$, we obtain

$$E_c(\mathbf{k}) = E_c(0) + \frac{\hbar|\mathbf{k}|^2}{2m_e} \left(1 + \frac{P^2}{E_g}\right). \qquad (7.24)$$

[19] See, e.g., E. O. Kane, *Journal of Physics and Chemistry of Solids* **6**, 236 (1958); E. O. Kane, in: *Semiconductors and Semimetals*, edited by R. K. Willardson and A. C. Beer (Academic, New York, 1966), Vol. 1; P. Y. Yu and M. Cardona, *Fundamentals of Semiconductors* (Springer, 1996); and I. Vurgaftman, J. R. Meyer, and L. R. Ram-Mohan, *Journal of Applied Physics* **89**, 5815 (2001).

[20] See Equation (5.95) for the definition of effective mass.

	$\frac{m^*(0)}{m_e}$	E_g (eV)	P^2 (eV)
InSb	0.014	0.235	23.3
InAs	0.026	0.417	21.5
GaSb	0.039	0.812	27.0
GaAs	0.067	1.519	28.8
GaN	0.150	3.299	25.0

Table 7.1 Band edge electron effective mass $m^*(0)$, zero-temperature band gap E_g, and Kane's interband momentum matrix element P^2 for representative III-V semiconductors; $m_e = 9.11 \times 10^{-31}$ kg. The values for GaN are for the zinc-blende type. Source: I. Vurgaftman *et al.*, *Journal of Applied Physics* **89**, 5815 (2001).

Here, $P^2 := 2|\langle u_{c0}|\hat{p}_x|u_{v0}\rangle|^2/m_e$ is known as Kane's interband momentum matrix element, whose values for representative direct-gap III-V semiconductors are listed in Table 7.1. Equation (7.24) allows us to obtain the following relationship between the conduction-band-edge effective mass[20] and the band gap:

$$\frac{m^*}{m_e} = \left(1 + \frac{P^2}{E_g}\right)^{-1}. \tag{7.25}$$

This relationship indicates that the effective mass increases with increasing band gap, a trend seen in Table 7.1.

7.4 Semiconductor Quantum Structures

Most III-V and II-VI compound semiconductors crystallize in the same form, as the zinc-blende structure, shown in Figure 7.11(b). Figure 7.15 plots different III-V semiconductors as a function of lattice constant (a) and band gap (E_g). Each compound is depicted as an open circle, with the lattice constant on the bottom axis and the band gap on the left vertical axis. A point on a line connecting two circles represents an alloy of certain compositions of the end members. For example, the line connecting GaAs and InAs represents $In_xGa_{1-x}As$ with $0 \leq x \leq 1$. The band gap and lattice constant of an alloy $In_xGa_{1-x}As$ can be approximated as the weighted average of the end values, i.e., $E_g(x) = xE_g^{InAs} + (1-x)E_g^{GaAs}$ and $a(x) = xa^{InAs} + (1-x)a^{GaAs}$.

Figure 7.15 The lattice constants (bottom axis) and band gaps (left axis) of representative III-V compound semiconductors. Each semiconductor is shown as an open circle. © Deyin Kong (Rice University).

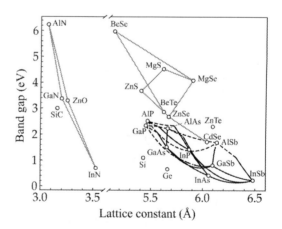

The most intriguing situation occurs when the line connecting two compounds is vertical. This happens for GaAs and AlAs; see Figure 7.15. In this case, these compounds have not only the same crystal structure but also the same lattice constant, i.e., $a(x) = a^{GaAs} = a^{AlAs}$, independently of x. In this sense, GaAs and AlAs are said to be *lattice-matched*. Also, InP and $In_{0.53}Ga_{0.47}As$ are lattice-matched, and InAs, GaSb, and AlSb are nearly lattice-matched. Lattice-matched semiconductors can be grown on top of each other to make heterostructures; this is known as epitaxial growth. In particular, using modern crystal growth techniques such as molecular beam epitaxy, one can make ultrathin layers with atomically smooth interfaces.

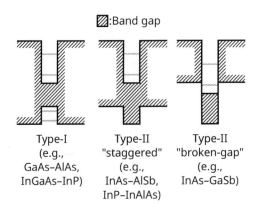

⬚ :Band gap

Type-I
(e.g.,
GaAs–AlAs,
InGaAs–InP)

Type-II
"staggered"
(e.g.,
InAs–AlSb,
InP–InAlAs)

Type-II
"broken-gap"
(e.g.,
InAs–GaSb)

Figure 7.16 There are three types of semiconductor quantum wells, depending on the band offsets of different semiconductors. © Deyin Kong (Rice University).

A quantum well arises when an ultrathin layer of a certain semiconductor is sandwiched between layers of another semiconductor with a different band gap. Depending on not only the band gap difference but also the electron affinity difference, three types of quantum wells exist; see Figure 7.16. In a Type-I quantum well, the band gap of the central layer (the "well") is completely energetically contained within the band gap of the outer layers (the "barriers"). Both electrons and holes are quantum mechanically confined, in a Type-I quantum well. In a Type-II *staggered* quantum well, the top of the valence band of the outer layers is higher than that of the inner layer, so there is no quantum confinement for holes, i.e., the barrier height is negative. In a Type-II *broken-gap* quantum well, the top of the valence band of the outer layers is even higher than the bottom of the conduction band of the center layer, resulting in an artificial semimetal, where electrons and holes coexist even at zero temperature.

Modern epitaxy techniques further allow the growth of various heterostructures; see Figure 7.17. For example, one can repeat a quantum well multiple times to grow multiple quantum wells or *superlattices*. One can even continuously vary the composition x to create a parabolic or triangular quantum well. Alternatively, a tunneling structure can be created to modify nonlinear I-V curves reflecting quantum electron dynamics. Ultimately, one can fully quantum-engineer a structure – including the energy levels and wavefunctions – with tailored properties for specific device applications.

Figure 7.17 Various types of semiconductor heterostructures. © Deyin Kong (Rice University).

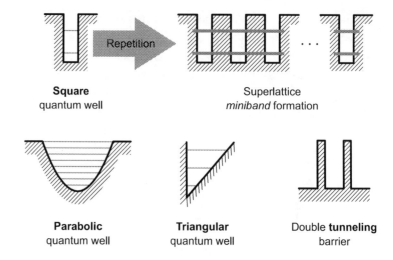

Square
quantum well

Superlattice
miniband formation

Parabolic
quantum well

Triangular
quantum well

Double **tunneling**
barrier

Finally, it should be mentioned that modern 2D materials such as graphene and transition metal dichalcogenides (see Figure 1.7) have expanded the realm of heterostructures. Since there is no need to connect adjacent layers through chemical bonding, these layered 2D materials can be placed on top of each other without lattice matching. The interlayer coupling is mediated via weak van der Waals interaction, and the resulting heterostructures are called van der Waals heterostructures. The absence of lattice-matching requirements enables heterostructures of arbitrary types, significantly widening the range of engineering possibilities of artificial quantum semiconductor structures with desired properties.

7.5 Graphene

Graphene, first successfully isolated in monolayer form from bulk graphite in the mid-2000s,[21] is the most prototypical 2D material. It is an extremely thin material, that is, just one atom thick. Electronic motion in the perpendicular direction is absolutely absent, while in-plane motion is characterized by extremely high mobilities. Note that graphene is a *natural* 2D system in the sense that one does not need lithographic techniques to fabricate it, unlike artificial quantum structures such as semiconductor quantum wells, wires, and dots. Graphene is 100% carbon (C), consisting of a honeycomb-like array of sp^2-bonded C atoms (see Figure 7.18). Namely, three of the four electrons of each C atom in the lattice are shared with the three neighboring C atoms to form strong covalent bonds (also known as σ bonds) through overlapping sp^2 orbitals. The remaining fourth electron, originally in the $2p_z$ orbital of each C atom, leads to the formation of delocalized states, known as π orbitals. These π electrons

[21] K. S. Novoselov *et al.*, *Science* **306**, 666 (2004); *Proceedings of National Academy of Science USA* **102**, 10451 (2005); *Nature* **438**, 197 (2005); Zhang *et al.*, *Nature* **438**, 201 (2005). For a review, see, e.g., A. K. Geim and K. S. Novoselov, "The rise of graphene," *Nature Materials* **6**, 183 (2007).

are responsible for electrical conduction in graphene.

Upon closer examination of the graphene crystal (Figure 7.18), we notice that there are two groups of atoms, or sublattices, A and B. Each sublattice is a triangular lattice whose lattice points can be represented as a linear combination of a_1 and a_2, known as the primitive lattice vectors. In the figure, all open-circle sites can be represented as $na_1 + ma_2$, where n and m are integers. Alternatively, one can consider this crystal as a single triangular lattice by associating two adjacent atoms (e.g., those enclosed by the dashed ellipse in the figure) and considering them as one basis unit. The interatomic bond length, a_{C-C}, is 0.142 nm, and

$$a := |a_1| = |a_2| = \sqrt{3}\, a_{C-C} = 0.246\,\text{nm}. \tag{7.26}$$

In Cartesian coordinates, a_1 and a_2 can be expressed as

$$a_1 = \frac{\sqrt{3}}{2} a\, e_x + \frac{1}{2} a\, e_y, \tag{7.27}$$

$$a_2 = \frac{\sqrt{3}}{2} a\, e_x - \frac{1}{2} a\, e_y. \tag{7.28}$$

The band structure of graphene has been known theoretically for a surprisingly long time. According to Wallace's seminal work in the 1940s,[22] the energy band dispersions for the π^* (conduction) and π (valence) bands of graphene within the nearest-neighbor tight-binding approximation are given by

$$E_{\pm}(k) = E_{2p} \pm \gamma_0 w(k), \tag{7.29}$$

where "+" and "−" refer to the conduction and valence bands, respectively, $k = (k_x, k_y)$ is the Bloch wavevector, $\gamma_0 \sim 3\,\text{eV}$ is a band parameter (known as the intersublattice transfer integral), E_{2p} is the energy of the carbon atomic $2p_z$ orbital, and

$$w(k) = \left[1 + 4\cos\left(\frac{\sqrt{3} k_x a}{2}\right) \cos\left(\frac{k_y a}{2}\right) + 4\cos^2\left(\frac{k_y a}{2}\right) \right]^{1/2}. \tag{7.30}$$

The reciprocal (k-space) lattice of graphene is shown in Figure 7.19. The primitive lattice vectors of the reciprocal lattice are[23]

$$b_1 = 2\pi \frac{a_2 \times e_z}{a_1 \cdot (a_2 \times e_z)} = \frac{2\pi}{a} \left(\frac{1}{\sqrt{3}} e_x + e_y \right), \tag{7.31}$$

$$b_2 = 2\pi \frac{e_z \times a_1}{a_2 \cdot (e_z \times a_1)} = \frac{2\pi}{a} \left(\frac{1}{\sqrt{3}} e_x - e_y \right). \tag{7.32}$$

The first Brillouin zone is a hexagon – see the center of the figure. Three of the six corners are the K points, with one of them located

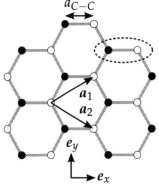

Figure 7.18 Graphene crystal structure. Each carbon (C) atom, represented by a solid or open circle, is connected to the three nearest-neighbor atoms through strong sp^2 covalent bonds. The C–C bond length $a_{C-C} = 0.142\,\text{nm}$. The open and solid circles belong to the two sublattices, the A and B sublattices, respectively.

[22] P. R. Wallace, *Physical Review* **71**, 622 (1947).

[23] See Section 7.2.

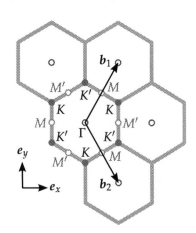

Figure 7.19 The reciprocal lattice of graphene. The first Brillouin zone is a hexagon, as shown by the solid gray outlines; b_1 and b_2 are the reciprocal lattice vectors, and Γ, K, K', M, and M' are high-symmetry points. The unit vectors in the x- and y-directions are defined as e_x and e_y, respectively. © Deyin Kong (Rice University).

[24] See, e.g., A. H. Castro Neto et al., "The eletronic properties of graphene," *Reviews of Modern Physics* **81**, 109 (2009).

at $k_K = (2\pi/\sqrt{3}a)e_x + (2\pi/3a)e_y$. The other three corners of the hexagon are the K' points, and one of them has coordinates $k_{K'} = (4\pi/3a)e_y$. The mid-points between K and K' are the M and M' points, and the origin $[(0,0)]$ is the Γ point as before.

From Equation (7.29), one can see that the energy separation between the conduction and valence bands, namely the band gap, is k-dependent. Specifically, since $E_g(k) := E_+(k) - E_-(k) = 2\gamma_0 w(k)$, we can deduce that

$$E_g(\Gamma) = 6\gamma_0 \sim 18 \text{ eV}, \tag{7.33}$$

$$E_g(M) = E_g(M') = 2\gamma_0 \sim 6 \text{ eV}, \tag{7.34}$$

$$E_g(K) = E_g(K') = 0 \text{ eV}. \tag{7.35}$$

Notably, the two bands touch each other exactly at the K and K' points. Therefore, graphene is a zero-gap material. Further examination of Equation (7.29) tells us the following:

(a) The bands are approximately parabolic near the Γ point.

(b) The bands are approximately linear near the K and K' points.

(c) The hexagon formed by connecting the M and M' points is an equi-energy line.

Point (b) above highlights the unusual photon-like, and thus massless, linear dispersion of electrons in graphene. The band structure of graphene is summarized in Figure 7.20, where the band dispersions and equi-energy contours are shown in detail. The massless and gapless band structure of graphene is truly unique, leading to a number of unusual properties.[24]

Figure 7.20 The band structure of graphene within the nearest-neighbor tight-binding approximation with $E_{2p} = 0$. (Upper) Energy dispersions $E = E(\boldsymbol{k})$ for the conduction and valence bands. (Lower) Equi-energy contours, $E = $ constant. © Deyin Kong (Rice University).

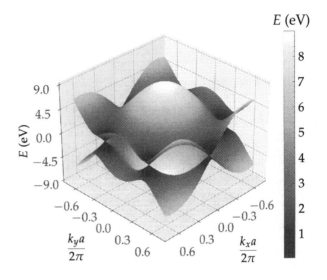

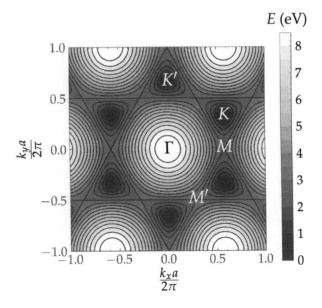

7.6   Carbon Nanotubes

Carbon nanotubes (CNTs) represent one of the most ideal realizations of a 1D material available today. They have extraordinarily large aspect ratios – nanometer-size diameters, with macroscopic lengths up to centimeters. Since their first identification and growth in the early 1990s, their unique physical, chemical, thermal, and mechanical properties have been extensively studied.[25] They have a thermal conductivity higher than any other materials, including diamond, and their mechanical strength (tensile strength and Young's modulus) is higher than that of steel. There are metallic and semiconducting CNTs: metallic CNTs conduct electricity like copper, and semiconducting CNTs emit light like GaAs with diameter-dependent wavelengths. Unlike graphene, semiconducting CNTs have band gaps and thus can be used to make transistors with large on/off ratios.

[25] For a review, see, e.g., S. Nanot et al., in: *Handbook of Nanomaterials*, edited by R. Vajtai (Springer, Berlin, 2013), pp. 105–146.

Figure 7.21 Crystal structure of SWCNTs. The chiral (or roll-up) vector, $C_h = n a_1 + m a_2$, connects two equivalent sites on the graphene sheet. An (n, m) SWCNT arises when the graphene sheet is rolled up so that the two ends of the vector C_h meet. The example here is for $(n, m) = (4, 2)$. In chiral nanotubes, $m \neq n, 0$. © Deyin Kong (Rice University).

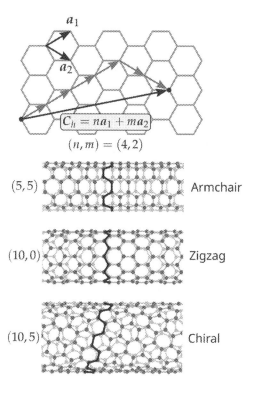

Structurally, a CNT, or more precisely, a *single-wall* CNT (SWCNT) can be viewed as a rolled-up graphene sheet in the form of a hollow cylinder. As shown in Figure 7.21, imagine a vector on the graphene sheet that connects two equivalent sites (within the same sublattice)

$$C_h = n a_1 + m a_2, \tag{7.36}$$

where n and m are integers and a_1 and a_2 are the primitive lattice vectors of graphene defined by Equations (7.27) and (7.28). This vector is referred to as the chiral (or roll-up) vector. An (n, m) SWCNT is formed when one cuts and rolls up the graphene sheet in such a way that C_h becomes the circumference of the resulting SWCNT. It immediately follows that the diameter of the (n, m) SWCNT, $d_t(n, m)$, is given by

$$d_t(n, m) = \frac{|C_h|}{\pi} = \frac{1}{\pi}\sqrt{n^2 a^2 + 2nm(a_1 \cdot a_2) + m^2 a^2}$$
$$= \frac{a}{\pi}\sqrt{n^2 + nm + m^2}. \tag{7.37}$$

For a $(10, 10)$ SWCNT, for example, the diameter is calculated to be $d_t(10, 10) = (a/\pi)(100 + 100 + 100)^{1/2} = 1.36\,\text{nm}$. The angle between C_h and a_1 is the chiral angle, denoted as α_{ch}. When $\alpha_{ch} = 0°$, we can immediately see that $m = 0$, and the nanotube is called a "zigazg" nanotube; see the zigzag pattern drawn along the circumference in Figure 7.21. When $\alpha_{ch} = 30°$, we deduce that $m = n$ (since the angle between a_1 and a_2 is $60°$); the nanotube is called an "armchair" nanotube. In all other cases (where $m \neq 0$ and $m \neq n$), the tubes are called "chiral" nanotubes. Structurally, these are the only three possible types of SWCNTs. In all three cases, the chiral angle can be generally calculated through

$$\alpha_{ch} = \tan^{-1}\left(\frac{\sqrt{3}m}{2n + m}\right). \tag{7.38}$$

One of the striking aspects of SWCNTs is their (n, m)-dependent electronic type, or metallicity. A slight change in (n, m), and thus in crystal structure, results in drastic changes in band structure, as shown in Figure 7.22. The $(10, 10)$ SWCNT is metallic, with no band gap, as shown in the energy versus density of states (DOS) graph.[26] The energy dispersions of the lowest-energy conduction and valence subbands are linear, and they touch each other, as in graphene. There is a higher-lying parabolic subband above the linear subband, both in the conduction and valence bands. In the $(10, 5)$ nanotube, on the other hand, there are only parabolic subbands, and there is a band gap separating the shaded (filled) valence subbands and the unshaded (unoccupied) conduction subbands. In both nanotubes, the DOS shows sharp peaks, the van Hove singularities characteristic of 1D materials.

The way in which the band structure of a SWCNT evolves out of that of graphene is analogous to the way in which a quantum wire (a 1D system) arises from a quantum well (a 2D system) through

[26] The density of states, $D(E)$, is defined such that $D(E)dE$ represents the number of states with energies between E and $E + dE$ available for electrons to occupy.

quantum confinement; see Figure 2.3. The graphene band structure depends on two continuous variables, (k_x, k_y), whereas that of a SWCNT contains a single continuous variable, k, which is the Bloch wavevector along the nanotube direction. The Bloch wavevector along the circumference is quantized in the following manner. Starting from a certain location on the tube, imagine moving around the circumference to return to the same point. Then the wavefunction must acquire the Bloch phase factor $e^{ik \cdot C_h}$, but since we are back to the original position, this phase factor must be 1, which is possible when $k \cdot C_h = 2\pi\mu$, where μ is an integer. Therefore, the energy E can be expressed in terms of k and μ, and each $E_\mu(k)$ is called a subband.

Figure 7.22 The crystal and band structure of $(10, 10)$ and $(10, 5)$ SWCNTs. In the density of states (DOS) of the $(10, 10)$ SWCNT, there is no band gap, and so this SWCNT is a metal. As in graphene, this system also has a linear dispersion, i.e., $E \propto k$. The $(10, 5)$ SWCNT is a semiconductor with a band gap, separating filled and empty subbands. In both SWCNTs, the DOS exhibits sharp peaks, known as van Hove singularities, characteristic of 1D systems. © Deyin Kong (Rice University).

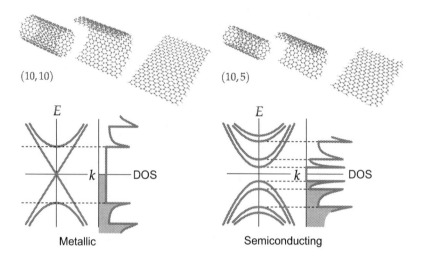

Whether a given (n, m) SWCNT is a metal or a semiconductor is determined by whether there is a subband that passes through the K or K' point of the original graphene reciprocal lattice, where the band gap is zero. If a subband includes the K point, its lowest-energy dispersion is massless ($E \propto k$) and gapless ($E_g = 0$) as in graphene. For example, in an armchair nanotube, where $n = m$ and thus $C_h = n(a_1 + a_2) = \sqrt{3}nae_x$, the quantization condition is $k \cdot C_h = \sqrt{3}nak_x = 2\pi\mu$.[27] Hence, the Bloch wavevector is given as $k = (2\pi\mu/\sqrt{3}na)e_x + ke_y$. This expression always includes $k_K = (2\pi/\sqrt{3}a)e_x + (2\pi/3a)e_y$ and $k_{K'} = (4\pi/3a)e_y$, so an armchair nanotube is a metal. A similar consideration for a general (n, m) SWCNT leads to the conclusion that the nanotube is metallic whenever $n - m$ is divisible by 3, i.e.,

[27] See Figure 7.19 for the definitions of e_x and e_y

$$(n - m) \bmod 3 = 0 \; \rightarrow \; \text{metallic}, \tag{7.39}$$

$$(n - m) \bmod 3 = \pm 1 \; \rightarrow \; \text{semiconducting}. \tag{7.40}$$

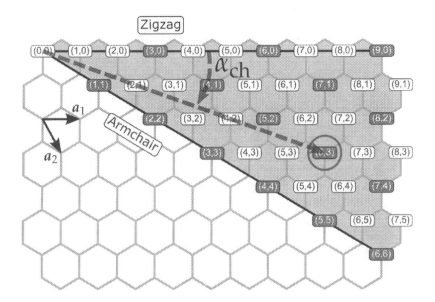

Figure 7.23 A variety of (n, m) carbon nanotubes exists depending on how the graphene sheet is rolled up. The chiral angle,

$$\alpha_{ch} = \tan^{-1}\left(\frac{\sqrt{3}m}{2n+m}\right),$$

is the angle between a_1 and the chiral vector $C_h = na_1 + ma_2$, and its value lies in the range between $0°$ (zigzag nanotubes) and $30°$ (armchair nanotubes). The length of C_h, $a(n^2 + nm + m^2)^{1/2}$, is the circumference of the corresponding nanotube. The (n, m) species, shown in dark boxes, are metals whereas those shown in white boxes are semiconductors. The C_h vector drawn here is for $(n, m) = (6, 3)$ as an example. © Deyin Kong (Rice University).

Using a zone-folding scheme within the nearest-neighbor tight-binding approximation, the band dispersions of an (n, m) SWCNT can be analytically obtained as

$$E_\mu(k) = E_{\text{graphene}}\left(k\frac{K_2}{|K_2|} + \mu K_1\right), \qquad (7.41)$$

where $E_{\text{graphene}}()$ is the graphene band dispersions given by Equation (7.29), $\mu = 0, 1, \ldots, N-1$ is the subband index, $N = 2(n^2 + nm + m^2)/d_R$ is the number of hexagons per unit cell of the (n, m) SWCNT, $d_R = \gcd(2m + n, 2n + m)$,[28] k is the Bloch wavevector along the 1D nanotube direction whose values lie within the first Brillouin zone defined as $[-\pi/Y, \pi/Y]$, $Y = |Y|$, $Y = t_1 a_1 + t_2 a_2$, $t_1 = (2m + n)/d_R$, $t_2 = -(2n + m)/d_R$, and K_1 and K_2 are the reciprocal lattice vectors of the SWCNT given by

$$K_1 = \frac{-t_2 b_1 + t_1 b_2}{N}, \qquad (7.42)$$

$$K_2 = \frac{mb_1 - nb_2}{N}, \qquad (7.43)$$

where b_1 and b_2 are the reciprocal lattice vectors of graphene, defined by Equations (7.31) and (7.32), respectively.

[28] gcd stands for the greatest common divisor.

Remark 7.1 Real-world applications of solid-state physics

The field of semiconductor research, or more generally solid-state physics research, is closely connected to device applications. The synergism that exists between device development and basic physics is truly exciting – any new phenomena observed in laboratories can be potentially utilized in devices in the future. For example, much of the present technology that fuels the information age is due to discoveries in solid-state research. In Table 7.1 we summarize key examples of this science–technology transition, primarily based on quantum-related phenomena discoveries and device-structure inventions.

Table 7.1: Major discoveries and inventions in solid-state technology.

Phenomenon/Structure	Discoverer(s)/Inventor(s)	Devices/Applications
Electroluminescence	H. J. Round[a]	LEDs,[b] indicators, lighting, communication, barcode scanners, computer mice
Superconductivity[c]	H. Kamerlingh Onnes	Superconducting magnets, MRI,[d] maglev trains
Transistor effect[e]	W. B. Shockley, J. Bardeen, and W. H. Brattain	Transistors, amplifiers, switches, integrated circuits, cell phones
The maser–laser principle[f]	C. H. Townes, N. G. Basov, and A. M. Prokhorov	Lasers, quantum electronics, barcode scanners, surgery, heat treatment
Quantum tunneling[g]	L. Esaki, I. Giaever, and B. D. Josephson	Tunnel diode, SQUIDs,[h] STM,[i] superconducting qubits
Charge coupled device[j]	W. S. Boyle and G. E. Smith	CCD[k] image sensors, cameras, digital imaging, spectrometers, astrophotography
Modulation doping[l]	R. Dingle, A. C. Gossard, and H. L. Störmer	HEMT,[m] millimeter-wave communication, radar, imaging, radio astronomy
The quantum Hall effect[n]	K. von Klitzing	Resistance standard, precision measurements of fundamental constants
High-temperature superconductivity[o]	J. G. Bednorz and K. A. Müller	Superconducting magnets, generators, motors, maglev trains
Integrated circuit[o]	J. S. Kilby	Personal computers, memories, CD players, cell phones
Semiconductor heterostructure[q]	Z. I. Alferov and H. Kroemer	Laser diodes, LEDs,[b] photodetectors, telecommunication, optoelectronics
Giant magnetoresistance[r]	A. Fert and P. Grünberg	Read-out heads for hard disks, information technology
Blue light emission from nitrides[s]	I. Akasaki, H. Amano, and S. Nakamura	Blue LEDs,[b] white-light LEDs, blue laser diodes, displays, lighting

[a]H. J. Round, *Electrical World* **19**, 309 (1907).

[b]LED: light-emitting diode.

[c]Nobel Prize in Physics in 1913 "for his investigations on the properties of matter at low temperatures which led, inter alia, to the production of liquid helium."[29]

[d]MRI: magnetic resonance imaging.

[e]Nobel Prize in Physics in 1956 "for their researches on semiconductors and their discovery of the transistor effect." See also a documentary video entitled "Transistor full documentary," available at https://www.youtube.com/watch?v=U4XknGqr3Bo&t=16s.

[f]Nobel Prize in Physics in 1964 "for fundamental work in the field of quantum electronics, which has led to the construction of oscillators and amplifiers based on the maser–laser principle."

[g]Nobel Prize in Physics in 1973 "for their discoveries regarding tunneling phenomena in solids."

[h]SQUID: superconducting quantum interference device.

[i]STM: scanning tunneling microscope.

[j]Nobel Prize in Physics in 2009 "for the invention of an imaging semiconductor circuit – the CCD sensor."

[k]CCD: charge coupled device.[30]

[l]US 4163237, Ray Dingle, Arthur Gossard, and Horst Störmer, "High mobility multilayered heterojunction devices employing modulated doping," 1979.

[m]HEMT: high electron mobility transistor.[31]

[n]Nobel Prize in Physics in 1985 "for the discovery of the quantized Hall effect."

[o]Nobel Prize in Physics in 1987 "for their important break-through in the discovery of superconductivity in ceramic materials."

[p]Nobel Prize in Physics in 2000 "for his part in the invention of the integrated circuit."

[q]Nobel Prize in Physics in 2000 "for developing semiconductor heterostructures used in high-speed- and opto-electronics."

[r]Nobel Prize in Physics in 2007 "for the discovery of giant magnetoresistance."

[s]Nobel Prize in Physics in 2014 "for the invention of efficient blue light-emitting diodes which has enabled bright and energy-saving white light sources."

[29] See "The discovery of superconductivity," *Physics Today* **63**, 38 (2010).

[30] W. S. Boyle and G. E. Smith, "Charge coupled semiconductor devices," *The Bell System Technical Journal* **49**, 587 (1970).

[31] For a review, see T. Mimura, "Development of high electron mobility transistor," *Japanese Journal of Applied Physics* **44**, 8263 (2005).

7.7 Chapter Summary

In this chapter, we learned about the electronic band structure of the real materials that are important in today's and tomorrow's quantum technology. By focusing on crystalline materials, to which the Bloch theorem applies, we introduced two distinct but complementary perspectives – the tight-binding picture (Section 7.1) and the nearly-free electron picture (Section 7.2). The tight-binding picture is useful for understanding the origins of bands as coming from atomic orbitals, while the nearly-free electron picture allows us to understand how a band gap arises as a result of Bragg diffraction. In those sections, we introduced such important concepts as the transfer integral, band widths, primitive lattice vectors, the Bragg and Laue conditions, Brillouin zones, and reciprocal lattices. Finally, we reviewed the band structure of various representative materials and their heterostructures, including both traditional technologically important materials and emerging low-dimensional materials with unique properties, such as graphene and carbon nanotubes. We noted that the quantum engineering of band structure and electronic and optical properties is possible through quantum confinement, alloying, and heterostructures.

7.8   Exercises

Exercise 7.1 (Valence bands in III-V semiconductors)

After taking into account spin–orbit coupling and using $|j, j_z\rangle = |\frac{3}{2}, \pm\frac{3}{2}\rangle$ and $|\frac{3}{2}, \pm\frac{1}{2}\rangle$, where j is the total angular momentum and j_z is its z component, as the basis functions, one can express the effective $\mathbf{k} \cdot \mathbf{p}$ Hamiltonian describing the highest four valence bands in III-V semiconductors as

$$
\mathbf{H}_{\mathrm{VB}} = \begin{bmatrix} H_\mathrm{h} & c & b & 0 \\ c^\dagger & H_\mathrm{l} & 0 & -b \\ b^\dagger & 0 & H_\mathrm{l} & c \\ 0 & -b^\dagger & c^\dagger & H_\mathrm{h} \end{bmatrix}, \tag{7.44}
$$

where

$$
H_\mathrm{h} = -\frac{\hbar^2}{2m_\mathrm{e}}(\gamma_1 + \gamma_2)(k_x^2 + k_y^2) - \frac{\hbar^2}{2m_\mathrm{e}}(\gamma_1 - 2\gamma_2)k_z^2,
$$

$$
H_\mathrm{l} = -\frac{\hbar^2}{2m_\mathrm{e}}(\gamma_1 - \gamma_2)(k_x^2 + k_y^2) - \frac{\hbar^2}{2m_\mathrm{e}}(\gamma_1 + 2\gamma_2)k_z^2,
$$

$$
b = -\frac{\hbar^2}{2m_\mathrm{e}}2\sqrt{3}\gamma_3(k_x - ik_y)k_z,
$$

$$
c = \frac{\hbar^2}{2m_\mathrm{e}}\sqrt{3}\left[\gamma_2(k_x^2 - k_y^2) - 2i\gamma_3 k_x k_y\right], \tag{7.45}
$$

and γ_1, γ_2, and γ_3 are material-dependent parameters usually referred to as Luttinger parameters.[32]

[32] J. M. Luttinger, *Physical Review* **102**, 1030 (1956).

(a) Diagonalize the Hamiltonian (7.44) to obtain the dispersions of the two doubly degenerate bands ("heavy-hole" and "light-hole" bands):

$$
E_\mathrm{v} = E_\mathrm{v}(0) + \frac{\hbar^2}{2m_\mathrm{e}} \times
$$
$$
\left[(1 - \gamma_1)k^2 \pm \sqrt{4\gamma_2^2 k^4 + 12(\gamma_3^2 - \gamma_2^2)(k_x^2 k_y^2 + k_y^2 k_z^2 + k_z^2 k_x^2)}\right]. \tag{7.46}
$$

(b) Plot the dispersions for the heavy- and light-hole bands for InP, whose Luttinger parameters are $\gamma_1 = 5.04$, $\gamma_2 = 1.55$, and $\gamma_3 = 2.4$, for the $[100]$, $[110]$, and $[111]$ orientations for $|\mathbf{k}| < 1\,\mathrm{nm}^{-1}$.

(c) Calculate the effective masses for the two bands m_{HH}^* and m_{LH}^* at the band edge ($k = 0$) for the three principal orientations ($[100]$, $[110]$, and $[111]$).

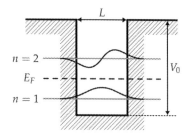

Inside the well:

$$\left[-\frac{\hbar^2}{2m_w^*}\frac{d^2}{dz^2}\right]\chi(z) = E\chi(z)$$

Outside the well:

$$\left[-\frac{\hbar^2}{2m_b^*}\frac{d^2}{dz^2} + V_0\right]\chi(z) = E\chi(z)$$

Figure 7.24 Quantum well with a finite barrier height for infrared photodetection. © Deyin Kong (Rice University).

Exercise 7.2 (Quantum well infrared photodetector)

Let us consider the single-quantum-well structure whose conduction band profile is depicted in Figure 7.24. The well is made up of a certain compound semiconductor AC, while the barriers are made from an alloy $A_{1-x}B_xC$, where $0 < x \le 1$. The goal is to construct a detector that resonantly absorbs electromagnetic radiation in the mid-infrared range at a wavelength of 10.6 µm utilizing the two lowest electron states in this structure with energies E_1 and E_2, respectively.

(a) Show that bound states in this structure can be found by solving the following equations:

$$\eta' = \gamma\xi\tan\xi, \tag{7.47}$$

$$\eta' = -\gamma\xi\cot\xi, \tag{7.48}$$

for even and odd solutions, respectively, together with

$$(\eta')^2 + \xi^2 = \frac{m_w^* V_0 L^2}{2\hbar^2}, \tag{7.49}$$

where m_w^* is the electron effective mass of the well material, m_b^* is the electron effective mass of the barrier material, V_0 is the barrier height, L is the well width, $\xi = (L/2\hbar)\sqrt{2m_w^* E}$, $\eta' = \eta\gamma$, $\eta = (L/2\hbar)\sqrt{2m_b^*(V_0 - E)}$, and $\gamma = \sqrt{m_w^*/m_b^*}$.

(b) Describe a way to solve the above equations graphically and derive the condition for having at least two bound states in the well.

(c) Assume that $m_w^* = 0.07m_e$, $m_b^*(x) = [0.07(1-x) + 1.05x]\,m_e$, and $V_0(x) = 0.3x\,\text{eV}$. Describe your strategy for finding the best values for x and L that will make the energy separation $E_2 - E_1$ equal the photon energy of 10.6 µm radiation.

Exercise 7.3 (Quantum engineering with 2D materials)

Refer to the solid lines (denoted as "PBE") in Figure 2(a) of J. Kang et al., "Band offsets and heterostructures of two-dimensional semiconductors," *Applied Physics Letters* **102**, 012111 (2013), when you answer the following questions.

(a) Propose an alloy of transition metal dichalcogenides (TMDs) whose band gap corresponds to a wavelength of 780 nm.

(b) Identify combinations of TMDs that can be used to construct Type-I and Type-II heterostructures.

(c) Which Type-II combination will provide the smallest (spatially) indirect band gap?

(d) Which combination of TMDs will create a conduction band quantum well with the largest barrier height?

Exercise 7.4 (Graphene band structure)

(a) Calculate the band gap E_g at the three high-symmetry points in k-space (the Γ, K, and M points) in units of γ_0.

(b) Draw the (dimensionless) band dispersions $[E(k) - E_{2p}]/\gamma_0$ along the Γ–K, Γ–M, and M–K lines in k-space.

(c) Show that the bands are approximately parabolic near the Γ point and approximately linear near the K point.

(d) Show that the hexagon formed by connecting the M and M' points is an equienergy line.

(e) Draw some equi-energy lines in k-space for the π^* band. Indicate the energy of each line (in units of γ_0 and measured from E_{2p}). Discuss how the shape of equi-energy lines changes with increasing energy around the K point.

Exercise 7.5 (Carbon nanotube band structure)

Consider the band structure of single-wall carbon nanotubes with chiral vector $C_h = na_1 + ma_2 = (n, m)$.

(a) Show the following relations: (i) $Y = \sqrt{3}|C_h|/d_R$, (ii) $|K_1| = 2/d_t$, and iii) $|K_2| = 2\pi/Y$.

(b) For $(3,2)$, $(4,3)$, $(6,2)$, and $(7,0)$ carbon nanotubes, calculate $|C_h|$, d_t, d_R, $Y = (t_1, t_2)$, $Y = |Y|$, and N, where N is the number of hexagons per unit cell (which equals the number of subbands for each of the conduction and valence bands).

(c) For the $(7,0)$ nanotube, draw all the cutting lines (line segments) representing the allowed 1D subbands in graphene k-space. Is this a metal or a semiconductor?

(d) Calculate and draw 1D band dispersions (E versus k) for the $(7,0)$ nanotube for $-\pi/Y < k < \pi/Y$, assuming $E_{2p} = 0$ and $\gamma_0 = 3\,\text{eV}$.

(e) From the band dispersions calculated above, determine the first five energy gaps: E_{11}, E_{22}, E_{33}, E_{44}, and E_{55}.

Appendix A
Classical Mechanics of Particles

ALL NONGRAVITATIONAL PHENOMENA OF CLASSICAL MECHANICS can in principle be predicted by quantum mechanics, but we have seen that the converse is not true. Therefore, it is not strictly necessary to learn classical physics before quantum physics. However, the philosophy of quantum mechanics is inherited from the philosophy of the more refined formulations of classical mechanics, and a strong intuition about classical philosophical approaches makes quantum mechanics much more understandable.

A.1 Hamilton's Principle

Newtonian mechanics postulates the existence of inertial frames, i.e., frames where Newton's equation of motion holds:

$$F = \frac{dp}{dt}. \tag{A.1}$$

If one has found an inertial frame and wishes to calculate the trajectories of n interacting particles in this frame, then one must solve

$$F_i = \frac{dp_i}{dt}, \qquad i = 1, 2, \ldots, n \tag{A.2}$$

given $3n$ initial positions and $3n$ initial velocities. In general, this is a system of $3n$ ordinary differential equations of second order, and is reasonably amenable to numerical methods. Though the Newtonian statement of mechanics is often sufficient (along with a good computer) to calculate dynamical trajectories, Hamilton's principle is another equivalent statement of mechanics that encapsulates the theory of mechanics in a single succinct equation. Before we can state Hamilton's principle, however, we need to introduce some language:

Definition A.1 *Let $q_1, q_2, \ldots, q_n$ denote the positions of a system of particles, and $\dot{q}_1, \dot{q}_2, \ldots, \dot{q}_n$ denote their velocities. The Lagrangian $\mathcal{L}$ of this system of particles is defined to be the difference between the kinetic energy, $\mathcal{T}$, and the potential energy, V, of all the particles in the system, i.e.,*

$$\mathcal{L}(q_1, \ldots, q_n; \dot{q}_1, \ldots, \dot{q}_n; t) := \mathcal{T} - V. \tag{A.3}$$

For example, a particle of mass m with kinetic energy $\frac{1}{2}m\dot{x}^2$ moving in a simple harmonic oscillator potential $V(x) = \frac{1}{2}\kappa_0 x^2$ has as its Lagrangian

$$\mathcal{L}(x, \dot{x}, t) = \frac{1}{2}m\dot{x}^2 - \frac{1}{2}\kappa_0 x^2. \tag{A.4}$$

We will generally try to restrict ourselves to the case of a single variable, simply because very little insight is gained but much notational difficulty is encountered in attempting to deal with the many-variable case. The Lagrangian (now in one variable) can be integrated to yield the action:

Definition A.2 *The action of a physical system is given by*

$$S[q] := \int_{t_1}^{t_2} \mathcal{L}(q, \dot{q}, t)\, dt. \tag{A.5}$$

The action measures the time-integrated difference between the kinetic and potential energies. The action S is an important example of a *functional* – a function that takes as its arguments functions, and returns a scalar quantity. Especially important for the study of quantum mechanics is the notion of a linear functional:

Definition A.3 (Linear functional) *Let V be a vector space, and J a functional. We say that J is a linear functional if and only if*

$$J[f + g] = J[f] + J[g], \qquad \forall f, g \in V \tag{A.6}$$

and

$$J[\alpha f] = \alpha J[f], \tag{A.7}$$

where α is a scalar quantity.

We can create a "calculus of functionals," similar to the standard calculus of variables, if we define a notion of the differentiation of a functional. This is called the variational derivative, and is defined as follows:

Definition A.4 (Variational derivative) *Let y and f be functions in a function space V, and let J be a functional. Then the variational derivative of J at y, denoted by δJ_y, is the linear functional such that*

$$J[y + f] - J[y] = \delta J_y[f] + \mathcal{O}(\|f\|^2). \tag{A.8}$$

The term "$\mathcal{O}(\|f\|^2)$" is to be understood as "an error of order $\|f\|^2$." For example, by Taylor series expansion, $e^x = 1 + x + \mathcal{O}(x^2)$.

We say that $\delta J_y = 0$ if and only if $\delta J_y[f] = 0$ for all $f \in V$. Similarly to ordinary calculus, $\delta J_y = 0$ indicates that y is an extremum of the functional J, either a maximum or a minimum.

With all this machinery in hand, we can now state Hamilton's principle:

Definition A.5 (Hamilton's principle) *Let q be any differentiable path that a particle could take while traveling between points $a = q(t_1)$ and $b = q(t_2)$. The actual path the particle takes extremizes the action, i.e., q is such that $\delta S_q = 0$.*

Using the Lagrangian given in Equation (A.4), the action is

$$S[x] = \int_{t_1}^{t_2} \left(\frac{1}{2} m\dot{x}^2 - \frac{1}{2} \kappa_0 x^2 \right) dt \tag{A.9}$$

When varying the action, we must vary it along allowed paths, i.e., the paths for which the endpoints remain at a and b. So, in

$$S[x+f] - S[x]$$

we must have $f(t_1) = f(t_1) = 0$, so that $(x+f)(t_1) = a$ and $(x+f)(t_2) = b$. Since

$$S[x+f] = \int_{t_1}^{t_2} \left[\frac{1}{2} m(\dot{x}+\dot{f})^2 - \frac{1}{2} \kappa_0 (x+f)^2 \right] dt \tag{A.10}$$

$$= S[x] + \int_{t_1}^{t_2} (m\dot{x}\dot{f} - \kappa_0 x f) dt + \int_{t_1}^{t_2} \left(\frac{1}{2} m\dot{f}^2 - \frac{1}{2}\kappa_0 f^2 \right) dt \tag{A.11}$$

then

$$S[x+f] - S[x] = \int_{t_1}^{t_2} (m\dot{x}\dot{f} - \kappa_0 x f) dt + \mathcal{O}(\|f\|^2).$$

Hence, the variation of S at x is

$$\delta S_x[f] = \int_{t_1}^{t_2} (m\dot{x}\dot{f} - \kappa_0 x f)\, dt. \tag{A.12}$$

Integrating the first term by parts and using the condition $f(t_1) = f(t_2) = 0$ gives

$$\delta S_x[f] = -\int_{t_1}^{t_2} (m\ddot{x} + \kappa_0 x) f\, dt. \tag{A.13}$$

Exercise A.1 *Show that the variation δS_x of Equation (A.13) is a linear functional.*

In order to find the extremes of S, we set $\delta S_x = 0$. A sufficient condition for this to occur is $m\ddot{x} + kx = 0$; in fact, it is also necessary. The proof of necessity is called the fundamental lemma of the calculus of variations:

Lemma A.1 (Fundamental lemma of the calculus of variations) *Let $g \in C[t_1, t_2]$. If*

$$\int_{t_1}^{t_2} g(t) f(t)\, dt = 0 \tag{A.14}$$

for all $f \in C[t_1, t_2]$ such that $f(t_1) = f(t_2) = 0$, then $g(t) \equiv 0$.

The proof of the fundamental lemma is simple but requires some background in analysis. With the lemma, we can say that $\delta S_x = 0$ if and only if $m\ddot{x} + kx = 0$, so that Hamilton's principle and Newton's laws are completely equivalent. However, going through all this work to get Newton's laws for a system we knew all about in introductory mechanics is a bit disappointing; it would be nice if Hamilton's principle was more powerful than Newton's laws. However, many regard Hamilton's principle as a particularly elegant statement of mechanics, perhaps because it anthropomorphizes nature. While Newton's laws tell you where a particle will go when subjected to forces, Hamilton's principle tells you *why* a particle moves as it does: the particle "wants" to extremize its action.

Exercise A.2 *Let $\mathcal{L} = \frac{1}{2}m\dot{x}^2$ be the Lagrangian of a free particle. Show that if x minimizes S then m must be positive.*

A.2 Lagrangian Mechanics

Newtonian mechanics is useful for computations, and Hamilton's principle is useful for gaining conceptual understanding. By use of variational calculus we can construct a statement of mechanics in between Hamilton's principle and Newtonian mechanics which lends itself to conceptual understanding and can also be used for numerical computations. This is Lagrangian mechanics, which we will derive from Hamilton's principle. First, consider the following multivariate Taylor expansion:

$$\mathcal{L}(q+f, \dot{q}+\dot{f}, t) = \mathcal{L}(q, \dot{q}, t) + f\frac{\partial \mathcal{L}}{\partial q} + \dot{f}\frac{\partial \mathcal{L}}{\partial \dot{q}} + \mathcal{O}(\|f\|^2). \tag{A.15}$$

Then we have

$$S[q+f] - S[q] = \int_{t_1}^{t_2} \left(f\frac{\partial \mathcal{L}}{\partial q} + \dot{f}\frac{\partial \mathcal{L}}{\partial \dot{q}} \right) dt + \mathcal{O}(\|f\|^2) \tag{A.16}$$

and if $f(t_1) = f(t_2) = 0$, integration by parts of the term involving $\dot{f}$ gives

$$S[q+f] - S[q] = \int_{t_1}^{t_2} \left[\frac{\partial \mathcal{L}}{\partial q} - \frac{d}{dt}\left(\frac{\partial \mathcal{L}}{\partial \dot{q}} \right) \right] f \, dt + \mathcal{O}(\|f\|^2). \tag{A.17}$$

So we take

$$\delta S_q[f] = \int_{t_1}^{t_2} \left[\frac{\partial \mathcal{L}}{\partial q} - \frac{d}{dt}\left(\frac{\partial \mathcal{L}}{\partial \dot{q}} \right) \right] f \, dt. \tag{A.18}$$

By use of the fundamental lemma, $\delta S_q = 0$ if and only if

$$\boxed{\frac{\partial \mathcal{L}}{\partial q} - \frac{d}{dt}\left(\frac{\partial \mathcal{L}}{\partial \dot{q}} \right) = 0}. \tag{A.19}$$

This is called Lagrange's equation of motion; it is a statement of mechanics completely equivalent to both Newton's laws and Hamilton's principle.

Exercise A.3 *Solve Equation (A.19) for the simple harmonic oscillator Lagrangian, Equation (A.4), given initial conditions $x(0) = A$ and $\dot{x}(0) = 0$.*

Exercise A.4 *Show that Lagrange's equation, Equation (A.19), is invariant under transformations of the form $\mathcal{L}' = \mathcal{L} + f(t)$.*

Exercise A.5 *Show that Lagrange's equation, Equation (A.19), is invariant under the transformation $\mathcal{L}' = \alpha\mathcal{L}$, where α is a scalar constant.*

Exercise A.6 *A cannonball is fired at time $t = 0$ from a point $(0,0)$ in the x–y plane, and lands at the point $(L, 0)$ at time t_f. It moves without friction under the influence of gravity. What is the Lagrangian for this system? Solve Lagrange's equation to obtain the path $(x(t), y(t))$ that the cannonball follows. For this path, what is the value of the action integral $S[x, y]$?*

A.3   Generalized Coordinates and Momenta

Lagrange's equation, Equation (A.19), contains certain dimensional quantities, such as $\partial\mathcal{L}/\partial q$ (with units of energy/length, or force), and $\partial\mathcal{L}/\partial\dot{q}$ (with units of energy/velocity, or momentum). If q is a Cartesian coordinate, such as x, then $\partial\mathcal{L}/\partial\dot{x} = p_x$, and $\partial\mathcal{L}/\partial x = -\dot{p}_x$. However, we have no stipulation that requires q to be a Cartesian coordinate with dimensions of length. For example, the position of a pendulum bob rotating around a pivot can be specified by the angle of deviation from the vertical, θ, rather than its x and y positions in the plane. The Lagrangian for a pendulum of fixed length ℓ with a bob having mass m is given by

$$\mathcal{L} = \frac{1}{2}m\ell^2\dot{\theta}^2 + mg\ell\cos\theta. \tag{A.20}$$

However, it is easy to see that $\partial\mathcal{L}/\partial\theta$ has units of energy, not force (as θ is dimensionless) and $\partial\mathcal{L}/\partial\dot{\theta}$ has units of energy $\times$ time (or action), and not units of momentum. However, the convenience associated with describing the position of a pendulum by a single dimensionless parameter rather than two parameters x and y with dimensions is such that we are compelled to retain the language associated with the case where the dimension of the coordinate q is length. Hence, we define the generalized momentum as

$$p := \frac{\partial\mathcal{L}}{\partial\dot{q}}, \tag{A.21}$$

even if the generalized coordinate q does not have units of length. Similarly, we define the generalized force by

$$F := \frac{\partial\mathcal{L}}{\partial q}. \tag{A.22}$$

Exercise A.7 *Verify that a pendulum of length ℓ and mass m moving about a frictionless pivot under the influence of gravity has Equation (A.20) as its Lagrangian if the potential energy at the pivot is zero. Show that Lagrange's equation, Equation (A.19), becomes the pendulum equation $\ddot{\theta} = -(g/\ell)\sin(\theta)$.*

A.4   Hamiltonian

The transition from Lagrangian mechanics to Hamiltonian mechanics resides in changing what we decide to regard as the fundamental objects. In Lagrangian mechanics, the fundamental objects are the generalized positions q and generalized velocities $\dot{q}$. In Hamiltonian mechanics, the fundamental objects are the generalized positions q and the generalized momenta p. Similarly to how we defined the Lagrangian before we constructed Lagrangian mechanics, we now define the Hamiltonian:

$$\mathcal{H}(q_1,\ldots,q_n;p_1,\ldots,p_n,t) := \sum_{i=1}^{n} p_i\dot{q}_i - \mathcal{L}(q_1,\ldots,q_n;\dot{q}_1,\ldots,\dot{q}_n;t). \tag{A.23}$$

In keeping with the philosophy of regarding the generalized momenta p_i and the generalized coordinates q_i as the fundamental objects, we regard each $\dot{q}_i$ as functions of these quantities; i.e., we write

$$\dot{q}_i = \dot{q}_i(q_1,\ldots,q_n;p_1,\ldots,p_n). \tag{A.24}$$

The Hamiltonian has an important property:

$$\frac{d\mathcal{H}}{dt} = \sum_{i=1}^{n}(\dot{p}_i\dot{q}_i + p_i\ddot{q}_i) - \sum_{i=1}^{n}\left(\frac{\partial\mathcal{L}}{\partial q_i}\dot{q}_i + \frac{\partial\mathcal{L}}{\partial\dot{q}_i}\ddot{q}_i\right) - \frac{\partial\mathcal{L}}{\partial t} = -\frac{\partial\mathcal{L}}{\partial t} \tag{A.25}$$

(which follows simply from Equation (A.21)). In the important case where $\partial\mathcal{L}/\partial t = 0$, $\mathcal{H}$ is constant in time.

In quantum mechanics, it is often sufficient to simply take $\mathcal{H} = \mathcal{T} + V$.

A.5 Hamilton's Equations of Motion

Just as we derived Lagrange's equation of motion from Hamilton's principle and the Lagrangian, we can derive what are known as Hamilton's equations from Hamilton's principle and the Hamiltonian, via the substitution

$$\mathcal{L}(q_1,\ldots,q_n;\dot{q}_1,\ldots,\dot{q}_n;t) = \sum_{i=1}^{n}p_i\dot{q}_i - \mathcal{H}. \tag{A.26}$$

To prevent notational pandemonium, we restrict to the case of a single generalized coordinate and write

$$\mathcal{L}(q,\dot{q},t) = p\dot{q} - \mathcal{H}(p,q,t). \tag{A.27}$$

Then the action is

$$S[q,p] = \int_{t_1}^{t_2}[p\dot{q} - \mathcal{H}(p,q,t)]dt \tag{A.28}$$

and for $C^1[t_1,t_2]$ functions (i.e., differentiable functions whose derivative is continuous) χ and η with $\chi(t_1) = \chi(t_2) = \eta(t_1) = \eta(t_2) = 0$, we have

$$S[q+\chi,p+\eta] = \int_{t_1}^{t_2}[(p+\eta)(\dot{q}+\dot{\chi}) - \mathcal{H}(p+\eta,q+\chi,t)]dt. \tag{A.29}$$

Let $\|(\eta,\chi)\| := \max\{\|\eta\|,\|\chi\|\}$. By a multivariate Taylor expansion,

$$\mathcal{H}(p+\eta,q+\chi,t) = \mathcal{H}(p,q,t) + \eta\frac{\partial\mathcal{H}}{\partial p} + \chi\frac{\partial\mathcal{H}}{\partial q} + \mathcal{O}(\|(\eta,\chi)\|^2). \tag{A.30}$$

Then

$$\begin{aligned}
S[q+\chi,p+\eta] &= \int_{t_1}^{t_2}\left(p\dot{q} + p\dot{\chi} + \dot{q}\eta - \mathcal{H}(p,q,t) - \eta\frac{\partial\mathcal{H}}{\partial p} - \chi\frac{\partial\mathcal{H}}{\partial q}\right)dt + \mathcal{O}(\|(\eta,\chi)\|^2)\\
&= S[q,p] + \int_{t_1}^{t_2}\left(p\dot{\chi} + \dot{q}\eta - \eta\frac{\partial\mathcal{H}}{\partial p} - \chi\frac{\partial\mathcal{H}}{\partial q}\right)dt + \mathcal{O}(\|(\eta,\chi)\|^2)\\
&= S[q,p] + \int_{t_1}^{t_2}\left(\left[-\frac{\partial\mathcal{H}}{\partial q} - \dot{p}\right]\chi + \left[\dot{q} - \frac{\partial\mathcal{H}}{\partial p}\right]\eta\right)dt + \mathcal{O}(\|(\eta,\chi)\|^2).
\end{aligned} \tag{A.31}$$

So we must take

$$\delta S_{q,p}[\chi,\eta] = \int_{t_1}^{t_2}\left(\left[-\frac{\partial\mathcal{H}}{\partial q} - \dot{p}\right]\chi + \left[\dot{q} - \frac{\partial\mathcal{H}}{\partial p}\right]\eta\right)dt, \tag{A.32}$$

which, by a straightforward extension of the fundamental lemma, is identically zero if and only if

$$\boxed{\dot{p} = -\frac{\partial \mathcal{H}}{\partial q}, \ \dot{q} = \frac{\partial \mathcal{H}}{\partial p}}. \tag{A.33}$$

These are called Hamilton's equations of motion.

Exercise A.8 *Write down the Hamiltonian for the simple harmonic oscillator (whose Lagrangian is given by Equation (A.4)). Write down and solve Equation (A.33) for this Hamiltonian, given $x(0) = A$ and $\dot{x}(0) = 0$.*

Exercise A.9 *Show that $\delta S_{q,p}$, Equation (A.32), is linear over the product space $C^1[t_1, t_2] \times C^1[t_1, t_2]$, i.e., if $(\chi, \eta) \mapsto \alpha(\xi, \eta)$, then $\delta S_{p,q}[\chi, \eta] \mapsto \alpha \delta S_{p,q}[\chi, \eta]$ and $\delta S_{q,p}[\chi_1 + \chi_2, \eta_1 + \eta_2] = \delta S_{q,p}[\chi_1, \eta_1] + \delta S_{q,p}[\chi_2, \eta_2]$.*

A.6 Poisson Brackets

Suppose that we have a system with n degrees of freedom, and let u and v be any infinitely differentiable functions of the generalized momenta $\{p_i\}_{i=1}^n$ and coordinates $\{q_i\}_{i=1}^n$ and of the time t. Then we can define the Poisson bracket of u and v as follows:

$$\{u, v\}_{\mathrm{PB}} := \sum_{i=1}^n \left(\frac{\partial u}{\partial q_i} \frac{\partial v}{\partial p_i} - \frac{\partial u}{\partial p_i} \frac{\partial v}{\partial q_i} \right). \tag{A.34}$$

Poisson brackets have interesting mathematical properties, including:

- bilinearity,

- skew-symmetry, i.e., $\{u, v\}_{\mathrm{PB}} = -\{v, u\}_{\mathrm{PB}}$, and

- the Jacobi identity, $\{u, \{v, w\}_{\mathrm{PB}}\}_{\mathrm{PB}} + \{v, \{w, u\}_{\mathrm{PB}}\}_{\mathrm{PB}} + \{w, \{u, v\}_{\mathrm{PB}}\}_{\mathrm{PB}} = 0$.

This means that the Poisson bracket is an example of a Lie bracket, which endows the space of infinitely differentiable functions with the structure of a Lie algebra.

Exercise A.10 *For a particle free to move in 3D space, compute the following Poisson brackets:*

(a) $\{x_i, x_j\}_{\mathrm{PB}}$ *(where $x_1 := x$, $x_2 := y$, $x_3 := z$).*

(b) $\{p_i, p_j\}_{\mathrm{PB}}$.

(c) $\{x_i, p_j\}_{\mathrm{PB}}$. Answer: δ_{ij}.

Exercise A.11 *Show that the Poisson bracket is bilinear.*

Exercise A.12 *Verify that Poisson brackets bear the following analogy to the product rule:*

$$\frac{d}{dx}(vw) = \frac{dv}{dx}w + v\frac{dw}{dx}, \tag{A.35}$$

$$\{u, vw\}_{\mathrm{PB}} = \{u, v\}_{\mathrm{PB}}w + v\{u, w\}_{\mathrm{PB}}. \tag{A.36}$$

For this reason, the map $\{u, \cdot\}_{\mathrm{PB}}$ is called a derivation.

Appendix B
Classical Theory of Waves

THE CONCEPT OF WAVE–PARTICLE DUALITY IS ONE OF THE CENTRAL IDEAS IN QUANTUM MECHANICS. As we saw in Chapter 2, particles such as electrons start exhibiting wave properties when they are confined in structures whose dimensions are comparable with or smaller than their wavelength (the de Broglie wavelength). Even though such *matter waves* obey the Schrödinger equation, which is distinctly different from the classical wave equation, they show basic wave phenomena such as interference and refraction, and many of the conventional wave quantities such as phase, period, frequency, wavelength, group velocity, etc., are fully meaningful and applicable. Here, we will review some of the basic concepts and phenomena of classical waves.

B.1 Wave Equation and Plane Waves

The wave equation obeyed by many classical waves – e.g., sound waves, vibrations of a string, and electromagnetic waves – is written as

$$\nabla^2 \Psi = \frac{1}{v^2} \frac{\partial^2 \Psi}{\partial t^2}. \tag{B.1}$$

This equation – a linear partial differential equation that is second order both in space, x, and time, t – supports waves that travel at speed v. Let us confine ourselves to 1D, where $\nabla^2 = \partial^2 / \partial x^2$. The general solutions to this equation (known as the d'Alembert solutions) are $f(x \pm vt)$, where f is an arbitrary function. The $+$ $(-)$ solution corresponds to a wave propagating to the left (right). We consider plane wave solutions in the form

$$\Psi(x,t) = A \exp\left[\pm ik(x \pm vt)\right], \tag{B.2}$$

where A and k are constants (the amplitude and wavenumber, respectively). Introducing angular frequency, $\omega := kv$, let us look closely at the right-propagating solutions, $\Psi(x,t) = A \exp\left[\pm i(kx - \omega t)\right]$, the real part of which is $\mathrm{Re}\,\Psi = A \cos\left(kx - \omega t\right)$. For a fixed position x this wave oscillates periodically in time with period $T = 2\pi/\omega$. On the other hand, for a fixed time t it oscillates periodically in space with period $\lambda = 2\pi/k$, which is also known as the wavelength. See Figure B.1.

We can also include the attenuation of wave amplitudes by introducing a complex wavenumber

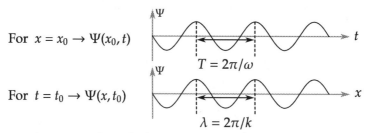

For $x = x_0 \rightarrow \Psi(x_0, t)$

$T = 2\pi/\omega$

For $t = t_0 \rightarrow \Psi(x, t_0)$

$\lambda = 2\pi/k$

Figure B.1 A monochromatic plane wave with angular frequency ω and wavenumber k. The corresponding temporal and spatial periods are T and λ, respectively, where λ is the wavelength. © Deyin Kong (Rice University).

$k = k' - ik''$. Using this complex k, we can write

$$\exp\left[-i(kx - \omega t)\right] = \exp\left[-i\left\{(k' - ik'')x - \omega t\right\}\right] = e^{-k''x} \exp\left[-i(k'x - \omega t)\right]. \tag{B.3}$$

The factor $e^{-k''x}$ shows that the amplitude decays exponentially with increasing distance x with a decay length of $\sim 1/k''$. In the case of a plane electromagnetic wave in a medium with complex refractive index $\tilde{N} = n_{\mathrm{op}} - i\kappa$, where n_{op} is the (real) refractive index and κ is the attenuation constant, we have $k = k_0 \tilde{N}$, where $k_0 = \omega/c$. Thus, $k'' = \kappa\omega/c$. The intensity of the wave $|\Psi|^2$ decays as $\propto e^{-2k''x} = e^{-2\kappa\omega x/c} = e^{-\alpha x}$, where $\alpha := 2\kappa\omega/c$ is the absorption coefficient.

B.2   Phase and Phase Velocity

Let us go back to $\Psi(x, t) = A \exp\left[-i(kx - \omega t)\right]$, which is a traveling wave, moving to the right. We also see that this is a *monochromatic* wave, having a well-defined angular frequency ω. We write

$$\phi(x, t) := kx - \omega t \tag{B.4}$$

as the phase of this traveling wave. In general, as can be seen in Equation (B.4), the phase ϕ varies with time. However, if we (as an observer) move at a particular constant velocity, the phase does *not* appear to be moving, and the shape of the wave looks stationary. This velocity, v_{ph}, is called the phase velocity. By requiring that

$$d\phi = \frac{\partial\phi}{\partial x}\,dx + \frac{\partial\phi}{\partial t}\,dt = k\,dx - \omega\,dt = 0, \tag{B.5}$$

we obtain

$$v_{\mathrm{ph}} = \frac{dx}{dt} = \frac{\omega}{k}. \tag{B.6}$$

For an electromagnetic wave traveling in vacuum, $v_{\mathrm{ph}} = c$, and thus, $\omega = ck$. In a medium, the dispersion relationship, $\omega = \omega(k)$, differs from this simple linear relation, and the phase velocity accordingly changes to $v_{\mathrm{ph}} = c' = \omega(k)/k$ (where k is real, assuming that the medium is lossless). By introducing the refractive index $n_{\mathrm{op}}(k) := c/c'$, we can also write $\omega(k) = ck/n_{\mathrm{op}}(k)$.

More generally, waves can have non-plane-wave forms:

$$\Psi(\mathbf{r}, t) = a(\mathbf{r}) \exp\left[i(\omega t - g(\mathbf{r}))\right]. \tag{B.7}$$

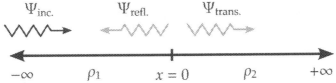

Figure B.2 An acoustic wave on a string is incident from $x = -\infty$. The string consists of two sections with different densities, ρ_1 (in region $x < 0$) and ρ_2 (in region $x > 0$). The wave is partially reflected and partially transmitted at $x = 0$. © Deyin Kong (Rice University).

Here, $g(\mathbf{r})$ is a real scalar function. The surface defined by

$$g(\mathbf{r}) = \text{constant} \tag{B.8}$$

is called the co-phasal surface (or wave surface), and the phase velocity in this general case is defined as

$$v_{\text{ph}}(\mathbf{r}) = \frac{\omega}{|\nabla g(\mathbf{r})|}. \tag{B.9}$$

This is the velocity with which the co-phasal surface advances. We can see that, if $g(\mathbf{r}) = kx$, then $|\nabla g(\mathbf{r})| = k$.

B.3 Reflection and Transmission

The string shown in Figure B.2 consists of two sections – a negative section ($x < 0$) with mass density ρ_1 and a positive section ($x > 0$) with mass density ρ_2. Although the angular frequency ω is constant in both sections, the velocity depends on the density as $v_1 = \sqrt{\mathbb{T}/\rho_1}$ and $v_2 = \sqrt{\mathbb{T}/\rho_2}$, where $\mathbb{T}$ is the tension. Accordingly, the wavenumber is $k_i = \omega/v_i = \omega\sqrt{\mathbb{T}/\rho_i}$ ($i = 1, 2$).

In the negative section, we have

$$\Psi_1(x, t) = \Psi_{\text{inc.}} + \Psi_{\text{refl.}} = Ae^{-i(k_1x - \omega t)} + Be^{-i(k_2x - \omega t)}, \tag{B.10}$$

and in the positive section, we have

$$\Psi_2(x, t) = \Psi_{\text{trans.}} = Ce^{-i(k_2x - \omega t)}. \tag{B.11}$$

Since the string is continuous, Ψ has to be continuous at $x = 0$:

$$\boxed{\Psi_1(0, t) = \Psi_2(0, t)}. \tag{B.12}$$

This is the first boundary condition. Suppose that the first derivative of the wave is discontinuous at $x = 0$, i.e., $\partial \Psi_1/\partial x|_{x=0} \neq \partial \Psi_2/\partial x|_{x=0}$. Then, $|\partial^2 \Psi/\partial x^2| = \infty$ at $x = 0$. However, this, together with the wave equation [Equation (B.1)], leads to $|\partial^2 \Psi/\partial t^2| = \infty$, which suggests infinite acceleration and thus is not allowed. Therefore, we obtain the other necessary boundary condition at $x = 0$:

$$\boxed{\left.\frac{\partial \Psi_1}{\partial x}\right|_{x=0} = \left.\frac{\partial \Psi_2}{\partial x}\right|_{x=0}}. \tag{B.13}$$

The two boundary conditions, Equations (B.12) and (B.13), lead to

$$A + B = C, \tag{B.14}$$

$$-k_1 A + k_1 B = -k_2 C. \tag{B.15}$$

By solving these coupled equations, we obtain

$$\frac{B}{A} = \frac{k_1 - k_2}{k_1 + k_2} = r, \tag{B.16}$$

$$\frac{C}{A} = \frac{2k_1}{k_1 + k_2} = t, \tag{B.17}$$

where r and t are reflection and transmission coefficients, respectively. For optical waves, since $k = n_{\mathrm{op}}\omega/c$, one can write

$$r = \frac{n_{\mathrm{op},1} - n_{\mathrm{op},2}}{n_{\mathrm{op},1} + n_{\mathrm{op},2}}, \tag{B.18}$$

$$t = \frac{2n_{\mathrm{op},1}}{n_{\mathrm{op},1} + n_{\mathrm{op},2}}. \tag{B.19}$$

B.4 Interference and Beats

Let us consider two wave sources, S_1 and S_2, which are located at different positions in space and are producing plane waves with the same amplitude and angular frequency, Ψ_0 and ω, respectively. These waves reach a certain position at time t with respective amplitudes

$$\Psi_1 = \Psi_0 e^{-i(\omega t + \phi_1)}, \tag{B.20}$$

$$\Psi_2 = \Psi_0 e^{-i(\omega t + \phi_2)}. \tag{B.21}$$

Here, $\phi_1 \neq \phi_2$ in general since the distance of this position from S_1 is in general different from the distance from S_2. Therefore, the total amplitude at this position at time t is the superposition of the two waves, $\Psi_3 = \Psi_1 + \Psi_2$, and its intensity is

$$|\Psi_3|^2 = |\Psi_1 + \Psi_2|^2 = (\Psi_1 + \Psi_2)(\Psi_1^* + \Psi_2^*) = |\Psi_1|^2 + |\Psi_2|^2 + \Psi_1 \Psi_2^* + \Psi_1^* \Psi_2. \tag{B.22}$$

The last two terms, $\Psi_1 \Psi_2^* + \Psi_1^* \Psi_2$, are the interference terms and can be modified to

$$\Psi_1 \Psi_2^* + \Psi_1^* \Psi_2 = \Psi_1 \Psi_2^* + (\Psi_1 \Psi_2^*)^* = 2\,\mathrm{Re}\,(\Psi_1 \Psi_2^*) = 2\,\mathrm{Re}\,\left(\Psi_0\, e^{-i(\omega t + \phi_1)}\, \Psi_0^*\, e^{i(\omega t + \phi_2)}\right)$$

$$= \mathrm{Re}\,\left(|\Psi_0|^2\, e^{-i(\phi_1 - \phi_2)}\right) = 2|\Psi_0|^2 \cos(\phi_1 - \phi_2). \tag{B.23}$$

Combining Equations (B.22) and (B.23) while noting that $|\Psi_1|^2 = |\Psi_2|^2 = |\Psi_0|^2$, we get

$$|\Psi_3|^2 = 2|\Psi_0|^2 [1 + \cos(\phi_1 - \phi_2)] = 4|\Psi_0|^2 \cos^2\left(\frac{\phi_1 - \phi_2}{2}\right). \tag{B.24}$$

When the two waves are "in phase," i.e., $\phi_1 = \phi_2$, we have $\cos^2\{(\phi_1 - \phi_2)/2\} = 1$ and thus $|\Psi_3|^2 = 4|\Psi_0|^2$, which is *constructive* interference. This looks as though $1 + 1 = 4$! On other other hand, when the two waves are "out of phase," i.e., $\phi_1 - \phi_2 = \pi$, we have $\cos^2\{(\phi_1 - \phi_2)/2\} = 0$ and thus $|\Psi_3|^2 = 0$, which is *destructive* interference. In this case, the sum of 1 and 1 appears to be zero.

Let us now look at a slightly different situation when $\phi_1 = \phi_2 = 0$ but the two frequencies ω_1 and ω_2 are different, i.e.,

$$\Psi_1 = \Psi_0\, e^{-i\omega_1 t}, \tag{B.25}$$

$$\Psi_2 = \Psi_0\, e^{-i\omega_2 t}. \tag{B.26}$$

In this case, the square of the absolute value of the sum of the two waves is

$$|\Psi_3|^2 = |\Psi_1 + \Psi_2|^2 = \left(\Psi_0\, e^{-i\omega_1 t} + \Psi_0\, e^{-i\omega_2 t}\right)\left(\Psi_0^*\, e^{i\omega_1 t} + \Psi_0^*\, e^{i\omega_2 t}\right)$$

$$= 2|\Psi_0|^2[1 + \cos{(\omega_1 - \omega_2)t}] = 4|\Psi_0|^2 \cos^2\left\{\frac{(\omega_1 - \omega_2)t}{2}\right\}. \tag{B.27}$$

Therefore, the wave intensity at this location *oscillates* in time at a frequency depending on the difference between the frequencies of the original two waves. This phenomenon is known as beats. Typically, the two original frequencies are close to each other, i.e., $\omega_1 \sim \omega_2$, and hence the $\cos^2\{(\omega_1 - \omega_2)t/2\}$ term produces a very slow oscillation, with familiar examples in acoustics and music. We can interpret beats as time-dependent interference: the two waves constructively (destructively) interfere with each other whenever $t = 2\pi n/(\omega_1 - \omega_2)$ $(t = (2n+1)\pi/(\omega_1 - \omega_2))$, where n is an integer.

B.5   Group Velocity, Wavepackets, and the Uncertainty Principle

Let us generalize the description of beats in the last section by including spatial variations of the two waves. Here, we are interested in how the total wave, $\Psi_3 = \Psi_1 + \Psi_2$, changes with t and x when

$$\Psi_1 = \Psi_0\, e^{i(k_1 x - \omega_1 t)}, \tag{B.28}$$

$$\Psi_2 = \Psi_0\, e^{i(k_2 x - \omega_2 t)}. \tag{B.29}$$

As before, the information on interference is contained in the real part:

$$\text{Re}\,(\Psi_3) = \text{Re}\left(\Psi_0\, e^{i(k_1 x - \omega_1 t)} + \Psi_0\, e^{i(k_2 x - \omega_2 t)}\right) = \Psi_0[\cos{(k_1 x - \omega_1 t)} + \cos{(k_2 x - \omega_2 t)}]$$

$$= 2\Psi_0\left(\frac{\Delta k\, x - \Delta\omega\, t}{2}\right)\cos{(\bar{k}x - \bar{\omega}t)}, \tag{B.30}$$

where $\Delta k := k_1 - k_2$, $\Delta\omega := \omega_1 - \omega_2$, $\bar{k} := (k_1 + k_2)/2$, and $\bar{\omega} := (\omega_1 + \omega_2)/2$. Here we have assumed that Ψ_0 is a real number for simplicity. Since $k_1 \sim k_2$ and $\omega_1 \sim \omega_2$, $\Delta k \sim \Delta\omega \sim 0$ and therefore the first cosine factor represents a slowly varying envelope, whereas the second cosine factor oscillates at the original frequencies, both in space and time, since $\bar{k} \sim k_1 \sim k_2$ and $\bar{\omega} \sim \omega_1 \sim \omega_2$. In AM radio transmission the latter represents the fast-oscillating "carrier wave," and the former is the slowly changing signal wave. The velocity that describes how fast the envelope moves is called the *group velocity*, v_g, in this two-wave case and can be calculated as follows:

$$d\Phi = \frac{\Delta k}{2}dx - \frac{\Delta\omega}{2}dt = 0 \quad \Rightarrow \quad v_g = \frac{dx}{dt} = \frac{\Delta\omega}{\Delta k}. \tag{B.31}$$

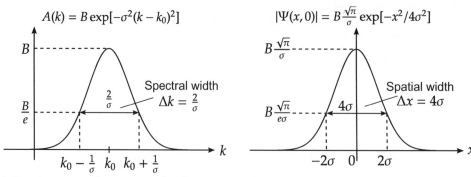

Figure B.3 A Gaussian wavepacket represented in k-space and real space with widths Δk and Δx, respectively. © Deyin Kong (Rice University).

More generally, the group velocity concept is meaningful and useful when there is a group of waves with similar frequencies, the general expression being given by

$$v_g = \frac{d\omega}{dk}. \tag{B.32}$$

When $A(k)$ is the distribution of waves in k, the total amplitude is given by the superposition

$$\Psi(x, t) = \int_{-\infty}^{\infty} A(k) e^{i(kx - \omega t)} dk. \tag{B.33}$$

This is a wavepacket, first introduced by Schrödinger. As an example, let us consider a Gaussian distribution,

$$A(k) = B \exp[-\sigma^2 (k - k_0)^2], \tag{B.34}$$

where σ is a positive constant; the width of this distribution, the spectral width, can be defined as $\Delta k = 2/\sigma$. This wavepacket moves in space at the group velocity. To estimate its width in x, let us consider the initial situation, $t = 0$:

$$\Psi(x, 0) = \int_{\infty}^{\infty} A(k) e^{ikx} dk = B \int_{-\infty}^{\infty} e^{-\sigma^2 (k-k_0)^2} e^{ikx} dk = B e^{ik_0 x} \int_{-\infty}^{\infty} e^{-\sigma^2 (k-k_0)^2} e^{i(k-k_0)x} dk$$

$$= B e^{ik_0 x} \int_{-\infty}^{\infty} e^{-\sigma^2 u^2} e^{iux} du = B \frac{\sqrt{\pi}}{\sigma} e^{ik_0 x} e^{-x^2/4\sigma^2}. \tag{B.35}$$

This is a Gaussian distribution in x with spatial width $\Delta x = 4\sigma$. Note that in deriving this result we used the following formula: $\int_{-\infty}^{\infty} e^{-ax^2} dx = \sqrt{\pi/a}$.

We can see that the product of the spectral width and the spatial width is constant, i.e.,

$$\Delta x \, \Delta k = 4\sigma \times \frac{2}{\sigma} = \text{constant}. \tag{B.36}$$

Thus, if the spectral bandwidth is large, then the wavepacket is spatially well confined. On the other hand, if the spectral distribution is narrow then the wavepacket is spatially spread. This evidently underlies Heisenberg's uncertainty principle, described in Section 3.5.2, if one recalls that $p = \hbar k$ for a free particle. In other words, if we have a well-defined single momentum then we completely lose any information on the position x, and vice versa. The group velocity can be well defined only when the wave can be satisfactorily represented by a wavepacket. If we include the time dependence of the wavepacket in the current problem, the wavepacket moves as a function of time, and the group velocity is the propagation velocity of the wavepacket.

B.6 Phase Velocity and Group Velocity

We have defined two velocities for the motions of waves, the phase velocity and the group velocity:

$$v_{\text{ph}} = \frac{\omega}{k}, \tag{B.37}$$

$$v_{\text{g}} = \frac{d\omega}{dk}. \tag{B.38}$$

The phase velocity is meaningful only for waves propagating with a well-defined angular frequency ω and wavenumber k. If the wave is inhomogeneous either in space or time, then the phase velocity is ill defined. The group velocity is meaningful only when there is a superposition of different waves with slightly different phase velocities.

It should be noted that the velocity at which any information is transmitted is the group velocity, not the phase velocity. This is important since, in some physical situations, $v_{\text{ph}} > c$ (where c is the speed of light) can happen, but that does not mean that one can transmit information faster than the light speed since $v_{\text{g}} > c$ never happens. Let us look at an example. For an electromagnetic wave with frequency ω propagating in a plasma with plasma frequency $\omega_{\text{p}}(< \omega)$, the refractive index can be written as

$$n_{\text{op}} = \sqrt{1 - \frac{\omega_{\text{p}}^2}{\omega^2}}. \tag{B.39}$$

The dispersion relation can then be written as

$$\omega = kc' = k\frac{c}{n_{\text{op}}} = \frac{kc}{\sqrt{1 - \omega_{\text{p}}^2/\omega^2}}. \tag{B.40}$$

It follows that the phase velocity is

$$v_{\text{ph}} = \frac{\omega}{k} = \frac{c}{\sqrt{1 - \omega_{\text{p}}^2/\omega^2}} > c. \tag{B.41}$$

Therefore, the phase velocity is greater than the speed of light. However, this does not mean that one can send signals faster than c. In fact, the group velocity is smaller than c, as shown below. By taking the square of both sides of Equation (B.40), we get

$$\omega^2 = \frac{k^2 c^2}{1 - \omega_{\text{p}}^2/\omega^2} \quad \Rightarrow \quad k = \frac{\sqrt{\omega^2 - \omega_{\text{p}}^2}}{c} \quad \Rightarrow \quad \frac{dk}{d\omega} = \frac{1}{c\sqrt{1 - \omega_{\text{p}}^2/\omega^2}}. \tag{B.42}$$

Therefore,

$$v_{\text{g}} = \frac{d\omega}{dk} = c\sqrt{1 - \frac{\omega_{\text{p}}^2}{\omega^2}} < c. \tag{B.43}$$

Appendix C
Electromagnetism: Maxwell's Equations

MAXWELL'S EQUATIONS are as follows:

$$\nabla \cdot \boldsymbol{D} = \rho, \text{ Coulomb's law} \tag{C.1}$$

$$\nabla \cdot \boldsymbol{B} = 0, \text{ No magnetic monopoles} \tag{C.2}$$

$$\nabla \times \boldsymbol{\mathcal{E}} = -\frac{\partial \boldsymbol{B}}{\partial t}, \text{ Faraday's law} \tag{C.3}$$

$$\nabla \times \boldsymbol{H} = \boldsymbol{J} + \frac{\partial \boldsymbol{D}}{\partial t}, \text{ Ampère's law} \tag{C.4}$$

These are are the most fundamental equations in electromagnetism, and accurately describe how electromagnetic fields behave and interact with matter. Even though they are classical equations, developed in the nineteenth-century pre-relativity and pre-quantum era, they have been shown to be valid both in the relativistic and quantum regimes without any corrections.

C.1 Definitions

In the above equations, $\nabla = (\partial/\partial x, \partial/\partial y, \partial/\partial z)$, $\boldsymbol{D}$ is the displacement field, $\boldsymbol{\mathcal{E}}$ is the electric field, ρ is the charge density, $\boldsymbol{B}$ is the magnetic field density, $\boldsymbol{H}$ is the magnetic field, $\boldsymbol{J} = \sigma \boldsymbol{\mathcal{E}}$ is the current density, and σ is the electrical conductivity. When an external electric field $\boldsymbol{\mathcal{E}}$ is applied to a material, an electric polarization field $\boldsymbol{\mathcal{P}} = \chi_e \varepsilon \boldsymbol{\mathcal{E}}$ is induced. Here, χ_e is the electric susceptibility of the material and $\varepsilon_0 = 8.85 \times 10^{-12}\,\text{Fm}^{-1}$ is the vacuum permittivity. The total field ($\boldsymbol{D}$) is the sum of the applied field ($\varepsilon_0 \boldsymbol{\mathcal{E}}$) and the induced field ($\boldsymbol{\mathcal{P}}$), i.e.,

$$\boldsymbol{D} = \varepsilon_0 \boldsymbol{\mathcal{E}} + \boldsymbol{\mathcal{P}} = \varepsilon_0 \boldsymbol{\mathcal{E}} + \chi_e \varepsilon_0 \boldsymbol{\mathcal{E}}$$
$$= \varepsilon_0 (1 + \chi_e) \boldsymbol{\mathcal{E}} = \varepsilon_0 \varepsilon_r \boldsymbol{\mathcal{E}} = \varepsilon \boldsymbol{\mathcal{E}}, \tag{C.5}$$

where $\varepsilon_r = 1 + \chi_e$ is the relative permittivity (or dielectric constant) of the material and is dimensionless. Similarly, when an external magnetic field $\boldsymbol{H}$ is applied to a material, a magnetization $\boldsymbol{M} = \chi_m \boldsymbol{H}$ is induced. Here, χ_m is the magnetic susceptibility of the material. The total field ($\boldsymbol{B}$) is the sum of the applied field ($\mu_0 H$) and the induced field ($\mu_0 M$)

$$B = \mu_0 H + \mu_0 M = \mu_0 H + \chi_m \mu_0 H$$
$$= \mu_0(1 + \chi_m)H = \mu_0\mu_r H = \mu H, \tag{C.6}$$

where $\mu_0 = 4\pi \times 10^{-7}\,\text{Hm}^{-1}$ is the vacuum permeability and $\mu_r = 1 + \chi_m$ is the relative permeability of the material and is dimensionless.

C.2 Wave Equation

First, let us assume that the medium is spatially homogeneous, so $\nabla \cdot D = \varepsilon\nabla \cdot \mathcal{E}$, and that there is no net charge, so $\rho = 0$. Then Coulomb's law [Equation (C.1)] becomes $\nabla \cdot \mathcal{E} = 0$. Next, we assume that the material is nonmagnetic ($\chi_m = 0$), so $\mu_r = 1$ and $\mu = \mu_0$. Further, using $B = \mu_0 H$ together with the facts that $D = \varepsilon\mathcal{E}$ and $J = \sigma\mathcal{E}$, we can rewrite Ampère's law [Equation (C.4)] as

$$\nabla \times B = \mu_0\sigma\mathcal{E} + \mu_0\varepsilon\frac{\partial\mathcal{E}}{\partial t}. \tag{C.7}$$

We have the following vector identity:

$$\nabla \times (\nabla \times \mathcal{E}) = \nabla(\nabla \cdot \mathcal{E}) - \nabla^2\mathcal{E} = -\nabla^2\mathcal{E}, \tag{C.8}$$

where we used Coulomb's law in the last step. By taking the rotation of both sides of Faraday's law [Equation (C.3)] and using the modified version of Ampère's law [Equation (C.7)], we get

$$\nabla \times (\nabla \times \mathcal{E}) = -\frac{\partial}{\partial t}(\nabla \times B) = -\mu_0\sigma\frac{\partial\mathcal{E}}{\partial t} - \mu_0\varepsilon\frac{\partial^2\mathcal{E}}{\partial t^2}. \tag{C.9}$$

By equating Equations (C.8) and (C.9), we obtain the wave equation for $\mathcal{E}$:

$$\boxed{\nabla^2\mathcal{E} = \mu_0\sigma\frac{\partial\mathcal{E}}{\partial t} + \mu_0\varepsilon\frac{\partial^2\mathcal{E}}{\partial t^2}}. \tag{C.10}$$

We seek a solution of plane wave form,

$$\mathcal{E} = \mathcal{E}_0\, e^{i(k\cdot r - \omega t)}, \tag{C.11}$$

where $k = (k_x, k_y, k_z)$ is the wavevector (or propagation vector) of the wave. This plane wave form allows us to replace ∇ with ik and $\partial/\partial t$ with $-i\omega$. With these replacements, we can rewrite the wave equation as $-k^2\mathcal{E} = -i\mu_0\sigma\omega\mathcal{E} - \mu_0\varepsilon\omega^2\mathcal{E}$. By equating the coefficient of $\mathcal{E}$ on both sides, we get the dispersion relationship

$$\boxed{k^2 = \mu_0\left(\varepsilon + i\frac{\sigma}{\omega}\right)\omega^2}. \tag{C.12}$$

In the simplest case, where the medium is a vacuum ($\varepsilon = \varepsilon_0$ and $\sigma = 0$), we recover the familiar linear dispersion relationship:

$$k^2 = \mu_0\varepsilon_0\omega^2 = \frac{\omega^2}{c^2} \quad \Rightarrow \quad k = k_0 = \frac{\omega}{c}. \tag{C.13}$$

More generally,

$$k = \sqrt{\mu_0 \left(\varepsilon + i\frac{\sigma}{\omega}\right)} \; \omega = \sqrt{\varepsilon_r + i\frac{\sigma}{\varepsilon_0\omega}} \; \frac{\omega}{c}. \tag{C.14}$$

Thus, by introducing the complex refractive index $\tilde{N}$ through $k = \tilde{N}k_0 = \tilde{N}\omega/c$ and $\tilde{N} = \sqrt{\tilde{\varepsilon}}$, where $\tilde{\varepsilon}$ is a generalized dielectric constant, we can write

$$\tilde{N} = \sqrt{\tilde{\varepsilon}} = \sqrt{\varepsilon_r + i\frac{\sigma}{\varepsilon_0\omega}}. \tag{C.15}$$

Dielectric materials are characterized by the condition $\varepsilon_r \gg i\sigma/\varepsilon\omega$, and thus, $\tilde{\varepsilon} \approx \varepsilon_r$. In metals, on the other hand, $i\sigma/\varepsilon_0\omega \gg \varepsilon_r$ and thus $\tilde{\varepsilon} \approx i\sigma/\varepsilon_0\omega$.

C.3   The Drude Model

Here, our goal is to obtain an expression for the electrical conductivity σ of a metal using classical mechanics. Specifically, we consider electrons in solids as classical free particles experiencing scattering from time to time. We introduce a phenomenological scattering time τ (or scattering rate τ^{-1}) as the average time of free movement between two scattering events. We apply an external DC electric field $\mathcal{E}$, which exerts an electrostatic force $F = -e\mathcal{E}$ on the electrons, and let $p(t)$ be the forward momentum of an electron at time t. During the period between t and $t + \Delta t$, the probability of scattering can be considered to be $\Delta t/\tau$, so a fraction $1 - \Delta t/\tau$ of electrons will not scatter as they accelerate and will contribute a change of momentum given by

$$\left(1 - \frac{\Delta t}{\tau}\right)[p(t) + F\Delta t + \mathcal{O}((\Delta t)^2)]. \tag{C.16}$$

On the other hand, a fraction $\Delta t/\tau$ of electrons do scatter and lose forward momentum, ending up in an equilibrium spatial distribution, so they contribute only

$$\frac{\Delta t}{\tau}[F\Delta t + \mathcal{O}((\Delta t)^2)] = \mathcal{O}((\Delta t)^2). \tag{C.17}$$

Therefore, the forward momentum at time $t + \Delta t$ can be written

$$\begin{aligned} p(t + \Delta t) &= \left(1 - \frac{\Delta t}{\tau}\right)[p(t) + F\Delta t] + \mathcal{O}((\Delta t)^2) \\ &= p(t) + \Delta t\left[-\frac{p(t)}{\tau} + F\right] + \mathcal{O}((\Delta t)^2), \end{aligned} \tag{C.18}$$

and we obtain the equation of motion

$$\frac{p(t + \Delta t) - p(t)}{\Delta t} \rightarrow \frac{dp(t)}{dt} = -\frac{p(t)}{\tau} + F. \tag{C.19}$$

Here, the first term on the right-hand side can be viewed as a viscous drag force (or velocity-dependent damping).

By introducing the mass m_e and average drift velocity v_D of the electrons via $p = m_e v_D$, and recalling that $F = -e\mathcal{E}$, we can rewrite the equation of motion as

$$\frac{dv_D}{dt} = -\frac{v_D}{\tau} - \frac{e\mathcal{E}}{m_e}. \tag{C.20}$$

In the steady state $dv_D/dt = 0$, and so we get

$$v_D = -\frac{e\tau}{m_e}\mathcal{E} = -\mu_e\mathcal{E}, \tag{C.21}$$

where μ_e is the electron mobility, which tells us how fast each electron moves in response to an electric field. If there are n_e electrons per unit volume, the total current density can be calculated to be

$$J = (-e)n_e v_D = \frac{n_e e^2 \tau}{m_e}\mathcal{E} = en_e\mu_e\mathcal{E}. \tag{C.22}$$

Since the conductivity σ is defined through $J = \sigma\mathcal{E}$, we obtain

$$\sigma_0 = \frac{n_e e^2 \tau}{m_e} = en_e\mu_e. \tag{C.23}$$

Here, the subscript 0 indicates that this is a DC conductivity. Equation (C.23) is known as the DC Drude conductivity. In a typical good conductor (e.g., copper, silver, or aluminum), $\sigma_0 \sim 10^7\,\mathrm{Sm^{-1}}$. Since n_e is typically $\sim 10^{29}\,\mathrm{m^{-3}}$ in these metals, the mobility and scattering time can be estimated to be $\mu_e = \sigma_0/en_e \sim 10^{-3}\,\mathrm{m^2\,V^{-1}s^{-1}}$ and $\tau = m_e\mu_e/e \sim 10\,\mathrm{fs}$, respectively.

Let us next consider the conductivity σ when the applied external electric field is an AC field, $\mathcal{E} \sim e^{-i\omega t}$. Assuming a linear response, the average drift velocity oscillates at the same frequency: $v_D \sim e^{-i\omega t}$. This means that we can replace d/dt with $-i\omega$ in Equation (C.20) and obtain

$$(-i\omega)v_D = -\frac{v_D}{\tau} - \frac{e\mathcal{E}}{m_e} \quad\Rightarrow\quad v_D = -\frac{(e\tau/m_e)}{1 - i\omega\tau}\mathcal{E}. \tag{C.24}$$

Therefore,

$$J = (-e)n_e v_D = \frac{(n_e e^2\tau/m_e)}{1 - i\omega\tau}\mathcal{E} = \frac{\sigma_0}{1 - i\omega\tau}\mathcal{E} = \sigma(\omega)\,\mathcal{E}.$$

$$\boxed{\sigma(\omega) = \frac{\sigma_0}{1 - i\omega\tau}}. \tag{C.25}$$

This is the AC Drude conductivity.

In the low-frequency, or DC, regime ($\omega\tau \ll 1$), the denominator of the AC Drude conductivity is approximately 1 and hence $\sigma(\omega) \simeq \sigma_0$. The condition $\omega\tau \ll 1$ can be written as $\omega \ll \tau^{-1}$. Physically, this means that the driving rate (ω) is much slower than the scattering rate (τ^{-1}), so the electrons do not notice that the applied field is oscillating. Note also that, in this regime, the conductivity is real, indicating that there is no phase lag between $\mathcal{E}$ and J. Since the Joule heating is proportional to $\mathcal{E} \cdot J$, there is maximum energy dissipation or absorption in this regime. In the high-frequency, or optical, regime ($\omega\tau \ll 1$), on the other hand, the conductivity can be approximated as

$$\sigma(\omega) \simeq i\frac{\sigma_0}{\omega\tau}. \tag{C.26}$$

Unlike the conductivity in the DC regime, this is now *purely* imaginary. Therefore, there is a 90° phase shift between J and $\mathcal{E}$, indicating that there is no energy loss in this regime. This is consistent with the fact that there is no free-carrier absorption in metals in the optical regime. Finally, in the intermediate regime ($\omega\tau \sim 1$), the conductivity is complex and can be written as $\sigma(\omega) = \sigma'(\omega) + i\sigma''(\omega)$, where

$$\sigma' = \frac{\sigma_0}{1 + (\omega\tau)^2}, \tag{C.27}$$

$$\sigma'' = \frac{(\omega\tau)\sigma_0}{1 + (\omega\tau)^2}. \tag{C.28}$$

These conductivity components are plotted in Figure C.1. Since typical values for τ^{-1} are $10^{12} - 10^{13}\,\mathrm{s}^{-1}$, the intermediate regime corresponds to the THz frequency range.

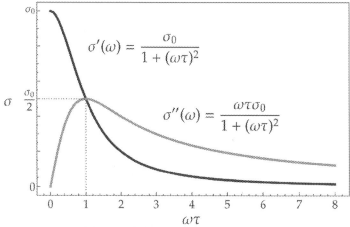

Figure C.1 The real and imaginary parts of the complex AC (or optical or dynamic) conductivity as a function of $\omega\tau$ obtained within the Drude model. © Deyin Kong (Rice University).

Finally, we examine how a metal responds to an electromagnetic wave in the optical (or high-frequency) regime, where $\omega\tau \gg 1$. We define the plasma frequency of the metal as

$$\omega_{\mathrm{p}} := \sqrt{\frac{n_e e^2}{\varepsilon m_e}} = \sqrt{\frac{\sigma_0}{\varepsilon_0 \varepsilon_{\mathrm{r}}\tau}}. \tag{C.29}$$

Then the AC Drude conductivity in the optical regime [Equation (C.26)] can be written as

$$\sigma(\omega) \simeq i\frac{\varepsilon_0 \varepsilon_{\mathrm{r}}\tau\omega_{\mathrm{p}}^2}{\omega\tau} = i\frac{\varepsilon_0 \varepsilon_{\mathrm{r}}\omega_{\mathrm{p}}^2}{\omega}. \tag{C.30}$$

The generalized dielectric constant for the metal can then be written as

$$\tilde{\varepsilon}(\omega) = \varepsilon_{\mathrm{r}} + i\frac{\sigma(\omega)}{\varepsilon_0\omega} = \varepsilon_{\mathrm{r}} - \frac{\varepsilon_{\mathrm{r}}\omega_{\mathrm{p}}^2}{\omega^2} = \varepsilon_{\mathrm{r}}\left(1 - \frac{\omega_{\mathrm{p}}^2}{\omega^2}\right). \tag{C.31}$$

This expression shows that when $\omega < \omega_{\mathrm{p}}$ the dielectric constant is negative. However, a negative dielectric constant leads to an imaginary refractive index since $\tilde{N} = \sqrt{\tilde{\varepsilon}}$. Furthermore, the general expression for the reflectivity of a material with index $\tilde{N} = n_{\mathrm{op}} - i\kappa$ is given by

$$\mathcal{R} = \frac{(1 - n_{\mathrm{op}})^2 + \kappa^2}{(1 + n_{\mathrm{op}})^2 + \kappa^2}; \tag{C.32}$$

a purely imaginary index (i.e., $n_{op} = 0$) results in total reflection, $\mathcal{R} = (1 + \kappa^2)/(1 + \kappa^2) = 1$. This perfect reflection for waves with $\omega < \omega_p$ is called plasma reflection, and the threshold $\omega = \omega_p$ is called the plasma edge and determines the characteristic color of a metal.

Appendix D
Parity

Let us consider the one-dimensional Schrödinger equation

$$\left(-\frac{\hbar^2}{2m}\frac{d^2}{dx^2} + V(x) \right) \psi(x) = E\psi(x), \tag{D.1}$$

where the potential energy $V(x)$ possesses inversion symmetry: $V(-x) = V(x)$. If we change x to $-x$ in Equation (D.1), it follows that

$$\left(-\frac{\hbar^2}{2m}\frac{d^2}{dx^2} + V(x) \right) \psi(-x) = E\psi(-x). \tag{D.2}$$

Hence, $\psi(-x)$ is an eigenfunction of the same Hamiltonian as $\psi(x)$ with the same eigenvalue. This indicates that $\psi(x)$ and $\psi(-x)$ are not independent of each other. Rather, they are proportional to each other, i.e.,

$$\psi(x) = \lambda\psi(-x), \tag{D.3}$$

where λ is a constant.

We introduce the parity operator $\hat{\Pi}$ as

$$\hat{\Pi}f(x) = f(-x), \tag{D.4}$$

where f is an arbitrary function. If we operate $\hat{\Pi}$ on $\psi(x)$ twice, we get

$$\hat{\Pi}^2\psi(x) = \hat{\Pi}[\hat{\Pi}\psi(x)] = \hat{\Pi}\psi(-x) = \hat{\Pi}[\lambda\psi(x)] = \lambda\hat{\Pi}\psi(x) = \lambda\psi(-x) = \lambda^2\psi(x). \tag{D.5}$$

However, by definition, operating $\hat{\Pi}$ on a function twice must return it to the original function, i.e., $\hat{\Pi}^2 = \hat{1}$, where $\hat{1}$ is the identity operator:

$$\hat{\Pi}^2\psi(x) = \hat{1}\psi(x) = \psi(x). \tag{D.6}$$

Equations (D.5) and (D.6) suggest that $\lambda^2 = 1$, or $\lambda = \pm 1$. When $\lambda = 1$,

$$\psi(x) = \psi(-x), \tag{D.7}$$

and the state is said to have *even* parity. When $\lambda = -1$,

$$\psi(x) = -\psi(-x), \tag{D.8}$$

and the state is said to have *odd* parity. In conclusion, if the potential energy has inversion symmetry $[V(x) = V(-x)]$, each eigenfunction must possess a definite parity, either even or odd.

Index

Printed in the United States
by Baker & Taylor Publisher Services